S.T.(P)
Technology
Today Series

A Series for Technicians

MATHEMATICS
FOR TECHNICIANS

Level II
Mechanical
Engineering
Mathematics

S.T.(P)
Technology
Today Series

A Series for Technicians *4127*

MATHEMATICS
FOR TECHNICIANS

Level II
Mechanical
Engineering
Mathematics

A. Greer
C.Eng., M.R.Ae.S.
Senior Lecturer
City of Gloucester College of Technology

G. W. Taylor
B.Sc.(Eng.), C.Eng., M.I.Mech.E.
Principal Lecturer
City of Gloucester College of Technology

Stanley Thornes (Publishers) Ltd.

First published in 1977 by:
Stanley Thornes (Publishers) Ltd
EDUCA House
32 Malmesbury Road
Kingsditch
CHELTENHAM GL51 9PL
England

Reprinted with minor correction 1978

ISBN 0 85950 054 3

Typeset in Monotype 10/12 Times New Roman by
Gloucester Typesetting Co Ltd
Printed and bound in Great Britain at
The Pitman Press, Bath

CONTENTS

AUTHORS' NOTE ON THE SERIES

Arising from the recommendations of the Haslegrave Report, the Technician Education Council has been set up. The Council has devised 'standard units', leading in various subjects to the award of its Certificates and Diplomas. The units are constructed at three levels, I, II and III.

A major change of emphasis in the educational approach adopted in T.E.C. Standard Units has been introduced by the use of 'objectives' throughout the courses, the intention being to allow student and lecturer to achieve planned progress through each unit on a step-by-step basis.

This set of books provides all the mathematics required for the T.E.C. Standard Units at each of the three levels. Each book follows a standard pattern, and each chapter opens with the words "After reaching the end of this chapter you should be able to:-" and this statement is followed by the objectives for that particular topic as laid down in the Standard Unit. Thereafter each chapter contains explanatory text, worked examples, and copious supplies of further exercises. As planned at present the series comprises:-

AN INTRODUCTORY COURSE   Level I (full unit)
MECHANICAL ENGINEERING
MATHEMATICS   Level II (half-unit)
PRACTICAL MATHEMATICS   Level II (half-unit)
ANALYTICAL MATHEMATICS   Level II (half-unit)
ELECTRICAL ENGINEERING
MATHEMATICS   Level II (full-unit)
MATHEMATICS FOR
ENGINEERING TECHNICIANS   Level III (full unit)

A. Greer
G. W. Taylor   Gloucester, 1977

1. TRIGONOMETRY

On reaching the end of this chapter you should be able to:-

1. *Define the secant, cosecant and cotangent ratios in terms of the sides of a right-angled triangle.*
2. *Determine, using tables, the secant, cosecant or cotangent of angles from 0° to 90° inclusive.*
3. *Derive the relationship* $\tan A = \dfrac{\sin A}{\cos A}$ *and* $\cot A = \dfrac{\cos A}{\sin A}$ *for a right-angled triangle.*
4. *Determine values for the six trigonometrical ratios for angles between 0° and 360°.*
5. *State the reciprocal relationships between the trigonometrical ratios for all angles.*
6. *Plot the graph of* $y = \sin A$, $y = \cos A$ *and* $y = \tan A$ *for angles between 0° and 360°.*
7. *Use the formulae for (i) sine rule (ii) cosine rule (iii) area of a triangle.*
8. *State the sine rule for a labelled triangle in the form* $\dfrac{a}{\sin A} = \dfrac{b}{\sin B} = \dfrac{c}{\sin C}$.
9. *Recognise conditions under which the sine rule can be used.*
10. *Apply the sine rule to the solution of practical problems.*
11. *State the cosine rule for a labelled triangle in the form* $a^2 = b^2 + c^2 - 2bc \cos A$.
12. *Recognise conditions under which the cosine rule can be used.*
13. *Apply the cosine rule to the solution of practical problems.*
14. *Calculate the area of a triangle using the formulae* $\frac{1}{2}ab \sin C$ *and* $\sqrt{s(s-a)(s-b)(s-c)}$.

TRIGONOMETRICAL RATIOS

Consider any angle θ which is bounded by the lines OA and OB as shown in Fig. 1.1. Take any point P on the boundary line OB. From P draw the line PM perpendicular to the other boundary line OA to meet OA at the point M. Then:

the ratio $\dfrac{MP}{OP}$ is called the *sine* of the angle AOB

the ratio $\dfrac{OM}{OP}$ is called the *cosine* of the angle AOB

and the ratio $\dfrac{MP}{OM}$ is called the *tangent* of the angle AOB

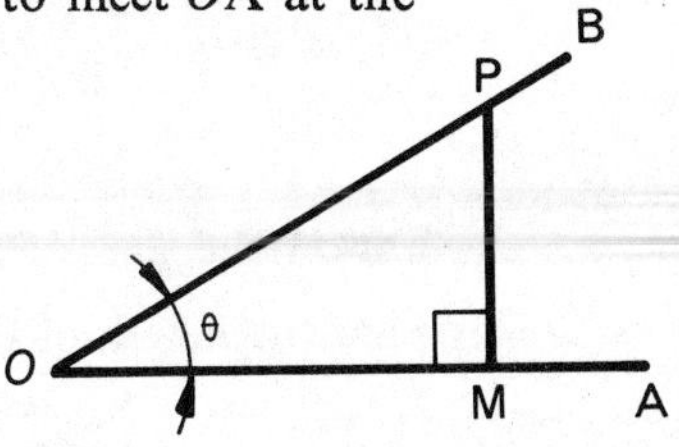

Fig. 1.1

THE SINE OF AN ANGLE

In any right-angled triangle (Fig. 1.2) the sine of an angle

$$= \frac{\text{side opposite the angle}}{\text{hypotenuse}}$$

$$\sin A = \frac{BC}{AC}$$

$$\sin C = \frac{AB}{AC}$$

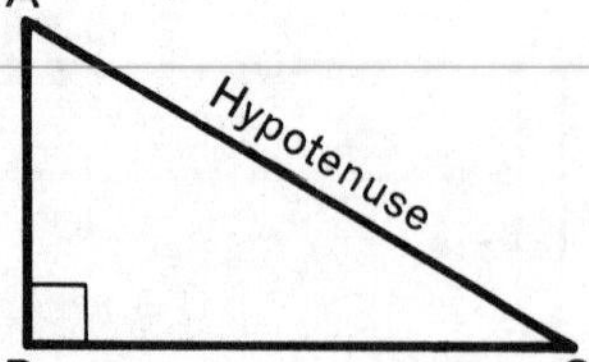

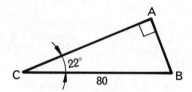

Fig. 1.2

The abbreviation 'sin' is usually used for 'sine'.

EXAMPLE 1

Find the length of the side AB in Fig. 1.3.

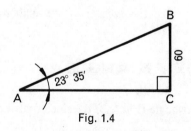

Fig. 1.3

AB is the side opposite to ∠ACB.
BC is the hypotenuse since it is opposite to the right-angle.

$$\frac{AB}{BC} = \sin 22°$$

$$AB = BC \times \sin 22° = 80 \times 0.374\,6 = 29.97 \text{ mm}$$

EXAMPLE 2

Find the length of the side AB in Fig. 1.4.

Fig. 1.4

BC is the side opposite to ∠BAC and AB is the hypotenuse.

$$\frac{BC}{AB} = \sin 23° 35'$$

$$BC = AB \times \sin 23° 35'$$

or $$AB = \frac{BC}{\sin 23° 35'} = \frac{60}{0.400\ 0} = 150\ \text{mm}$$

EXAMPLE 3

Find the angles CAB and ABC in the triangle ABC which is shown in Fig. 1.5.

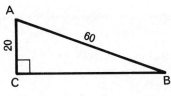

Fig. 1.5

$$\sin B = \frac{AC}{AB} = \frac{20}{60} = 0.333\ 3$$

From the sine tables B = 19° 28'

$$A = 90° - 19° 28' = 70° 32'$$

THE COSINE OF AN ANGLE

In any right-angled triangle (Fig. 1.6)

$$\text{the cosine of an angle} = \frac{\text{side adjacent to the angle}}{\text{hypotenuse}}$$

$$\cos A = \frac{AB}{AC}$$

$$\cos C = \frac{BC}{AC}$$

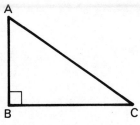

The abbreviation 'cos' is usually used for 'cosine'.

Fig. 1.6

EXAMPLE 4

Find the length of the side BC in Fig. 1.7.

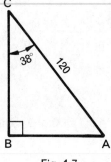

Fig. 1.7

Now BC is the side adjacent to $\angle$ BCA and AC is the hypotenuse.

$\therefore \qquad \dfrac{BC}{AC} = \cos 38°$

$BC = AC \times \cos 38° = 120 \times 0.788\,0 = 94.56 \text{ mm}$

EXAMPLE 5

Find the length of the side AC in Fig. 1.8.

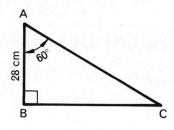

Fig. 1.8

$\dfrac{AB}{AC} = \cos 60°$

$AB = AC \times \cos 60°$

$AC = \dfrac{AB}{\cos 60°} = \dfrac{28}{0.500\,0} = 56 \text{ cm}$

EXAMPLE 6

Find the angle θ shown in Fig. 1.9.

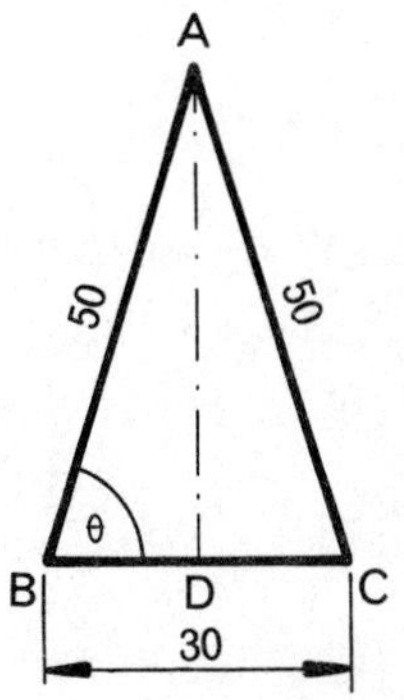

A

Fig. 1.9

50 50

θ

B| D |C

30

Since triangle ABC is isosceles the perpendicular AD bisects the base BC and hence BD = 15 mm

∴ $$\cos \theta = \frac{BD}{AB} = \frac{15}{50} = 0.3$$

$$\theta = 72° \, 32'$$

THE TANGENT OF AN ANGLE

In any right-angled triangle (Fig. 1.10),

the tangent of an angle $= \dfrac{\text{side opposite to the angle}}{\text{side adjacent to the angle}}$

$$\tan A = \frac{BC}{AB}$$

$$\tan C = \frac{AB}{BC}$$

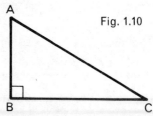

Fig. 1.10

The abbreviation 'tan' is usually used for 'tangent'.

EXAMPLE 7

Find the length of the side AB in Fig. 1.11.

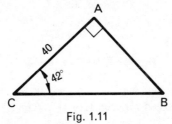

Fig. 1.11

$$\tan 42° = \frac{AB}{40}$$

$$AB = 40 \times \tan 42° = 40 \times 0.900\,4 = 36.02 \text{ mm}$$

EXAMPLE 8

Find the length of the side BC in Fig. 1.12.

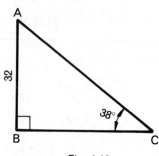

Fig. 1.12

There are two ways of doing this problem.

a)
$$\frac{AB}{BC} = \tan 38° \quad \text{or} \quad BC = \frac{AB}{\tan 38°}$$

∴
$$BC = \frac{32}{0.781\,3} = 40.96 \text{ mm}$$

b)   Since $C = 38°$, $A = 90° - 38° = 52°$

Now
$$\frac{BC}{AB} = \tan 52°$$
$$BC = AB \times \tan 52° = 32 \times 1.280 = 40.96 \text{ mm}$$

Both methods produce the same answer but method (**b**) is better because it is quicker and more convenient to multiply than to divide. Wherever possible the ratios should be arranged so that the quantity to be found is the numerator of the ratio.

Exercise 1

1)   Find the lengths of the sides marked x in Fig. 1.13, the triangles being right angled.

2)   Find the angles marked θ in Fig. 1.14, the triangles being right angled.

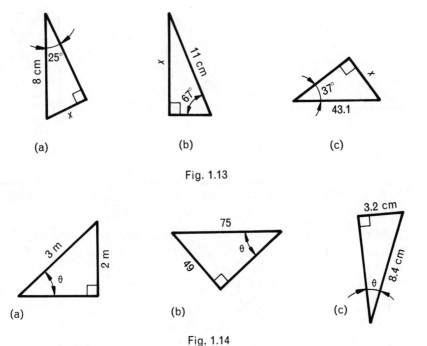

Fig. 1.13

Fig. 1.14

3) An equilateral triangle has an altitude of 18.7 cm. Find the length of the equal sides.

4) Find the altitude of an isosceles triangle whose vertex angle is 38° and whose equal sides are 7.9 m long.

5) The equal sides of an isosceles triangle are each 27 cm long, and the altitude is 19 cm. Find the angles of the triangle.

6) An isosceles trapezium is shown in Fig. 1.15. Find the length of the equal sides.

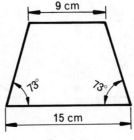

Fig. 1.15

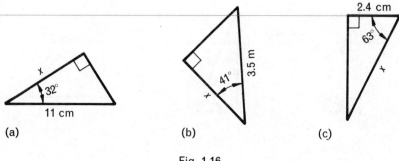

Fig. 1.16

7) Find the lengths of the sides marked x in Fig. 1.16 the triangles being right angled.

8) Find the angles marked θ in Fig. 1.17, the triangles being right angled.

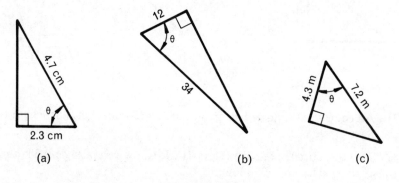

Fig. 1.17

9) An isosceles triangle has a base of 3.4 cm and the equal sides are each 4.2 cm long. Find the angles and the altitude of the triangle.

10) In Fig. 1.18 calculate $\angle$ BAC and the length BC.

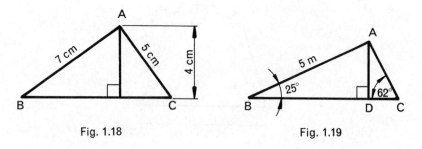

Fig. 1.18   Fig. 1.19

11) In Fig. 1.19 calculate:

   (a) BD   (b) AD   (c) AC   (d) BC

12) Find the lengths of the sides marked y in Fig. 1.20, the triangles being right angled.

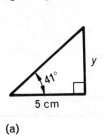

(a)

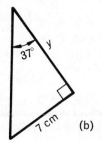

(b)

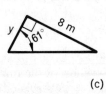

(c)

Fig. 1.20

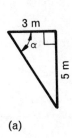

(a)

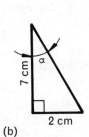

(b)

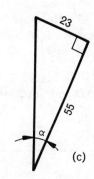

(c)

Fig. 1.21

13) Find the angles marked α in Fig. 1.21, the triangles being right angled.

14) An isosceles triangle has a base 10 cm long and the two equal angles are each 57°. Calculate the height of the triangle.

15) Calculate the distance l in Fig. 1.22.

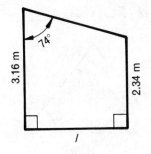

Fig. 1.22

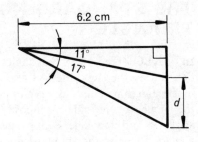

Fig. 1.23

16)   Calculate the distance d in Fig. 1.23.

17)   A vertical aerial, AC, has a wire stay, CD, 20 m long. The wire makes an angle of 50° with the ground. Find:

(a) the distance of the wire from the base of the pole;
(b) the height of the pole.

18)   From a point O, a run of pylons goes 20 km due west, 8 km due south and then 12 km due east, finally reaching a point A. Find the distance and bearing from O to A.

19)   A surveyor standing 50 m from a tower measures the angles of elevation of the top and bottom of a flagstaff on top of the tower as 55° and 51°. Calculate the height of the flagstaff.

20)   To find the height of a tower a surveyor stands some distance away from its base and finds that the angle of elevation to the top of the tower is 40°. He moves 100 m nearer to the base and finds that the angle of elevation is 65°. Find the height of the tower, assuming the ground to be horizontal.

21)   In Fig. 1.24, AB = 10 m, CD = 4 m and BC = 8 m. Find the angles BAD and CAD and also the length AD.

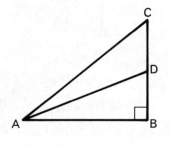

Fig. 1.24

THE STANDARD NOTATION FOR A TRIANGLE

In △ABC (Fig. 1.25) the angles are denoted by the capital letters as shown in the diagram. The side a lies opposite the angle A, the side b opposite the angle B and the side c opposite the angle C. This is the standard notation for a triangle.

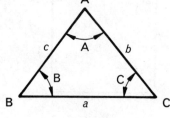

Fig. 1.25

RECIPROCAL RATIOS

In addition to sin, cos and tan there are three other ratios that may be obtained from a right angled triangle. These are:

cosecant (called cosec for short)

secant   (called sec for short)

cotangent (called cot for short)

Using the right angled triangle in Fig. 1.26:

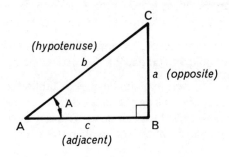

Fig. 1.26

$$\operatorname{cosec} A = \frac{\text{hypotenuse}}{\text{opposite}} = \frac{b}{a}$$

but it is already known that $\sin A = \dfrac{\text{opposite}}{\text{hypotenuse}} = \dfrac{a}{b}$

$\therefore$
$$\operatorname{cosec} A = \frac{1}{\sin A}$$

Similarly,

$$\sec A = \frac{\text{hypotenuse}}{\text{adjacent}} = \frac{b}{c}$$

but it is already known that $\cos A = \dfrac{\text{adjacent}}{\text{hypotenuse}} = \dfrac{c}{b}$

$\therefore$
$$\sec A = \frac{1}{\cos A}$$

and also,

$$\cot A = \frac{\text{adjacent}}{\text{opposite}} = \frac{c}{a}$$

but it is already known that $\quad \tan A = \dfrac{\text{opposite}}{\text{adjacent}} = \dfrac{a}{c}$

$\therefore \qquad\qquad\qquad\qquad \cot A = \dfrac{1}{\tan A}$

The reciprocal of x is $\dfrac{1}{x}$ and it may therefore be seen why the terms cosec, sec and cot are called 'reciprocal ratios', since they are equal respectively to $\dfrac{1}{\sin}$, $\dfrac{1}{\cos}$ and $\dfrac{1}{\tan}$.

It often helps to simplify calculations if the unknown length in a triangle trigonometry problem is made the numerator of a ratio, and the use of cosec, sec and cot in addition to sin, cos and tan makes this possible.

Tables of values for cosec, sec and cot of angles from 0° to 90° are usually included in books of standard mathematical tables.

These tables are used in a similar way as tables for sin, cos and tan; care must be taken when using the mean differences as to whether the tables give instructions to 'subtract' instead of adding as would normally be done. The examples which follow will explain this procedure.

EXAMPLE 9

Find the length of side x in Fig. 1.27.

Now $\qquad\qquad \dfrac{x}{35} = \operatorname{cosec} 35° \, 7'$

$\therefore \qquad\qquad x = 35 \times \operatorname{cosec} 35° \, 7'$

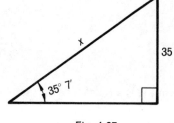

Fig. 1.27

From the table of cosecants, cosec 35° 6′ is read directly as 1.739 1. Looking in the mean difference column headed 1′ the value 7 (representing 0.000 7) is found. The instruction at the head of the table of cosecants indicates that the mean differences must be subtracted:

$\therefore \qquad \operatorname{cosec} 35° \, 7' = 1.739\,1 - 0.000\,7 = 1.738\,4$

and hence from above,

$$x = 35 \times 1.738\,4$$

$$= 60.844 \text{ mm}$$

EXAMPLE 10

Find the length of side y in Fig. 1.28.

$$\text{Now } \frac{y}{30} = \sec 64° \, 26'$$

$$\therefore \quad y = 30 \times \sec 64° \, 26'$$

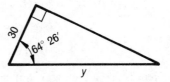

Fig. 1.28

From the table of secants, sec 64° 24′ is read directly as 2.314 4. Looking in the mean difference column headed 2′ the value 28 (representing 0.002 8) is found. Because there is no note to the contrary this value will be added to 2.314 4:

$$\therefore \qquad \sec 64° \, 26' = 2.314\,4 + 0.002\,8 = 2.317\,2$$

and hence from above,

$$y = 30 \times 2.317\,2$$

$$= 69.516 \text{ mm}$$

LOGARITHMS OF RECIPROCAL RATIOS

These tables are used to find the logarithm of a trigonometrical ratio in the same way as finding the ratio itself. They save the necessity of looking up the value of the ratio and then finding the corresponding logarithm.

EXAMPLE 11

Find the length of side b in Fig. 1.29.

$$\text{Now } \frac{b}{6.91} = \cot 39° \, 40'$$

$$\therefore \qquad b = 6.91 \times \cot 39° \, 40'$$

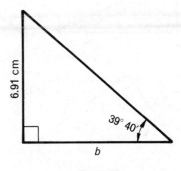

Fig. 1.29

The instruction at the head of the table of log cotangents indicates that the mean differences have to be subtracted.

	Number	log
$\therefore$ log cot 39° 40′ = log cot (39° 36′+4′)	6.91	0.839 5
= 0.082 3 − 0.001 0	cot 39° 40′	0.081 3
= 0.081 3	8.333	0.920 8

Hence, using the logarithm calculations shown in the table,

$$b = 8.333 \text{ cm.}$$

Exercise 2

Simple trigonometrical calculations involving the use of cosecant, secant and cotangent.

1) From the tables find the following:

(a) cosec 39° 27′   (b) cosec 67° 23′   (c) sec 11° 7′
(d) sec 49° 28′   (e) cot 37° 49′   (f) cot 74° 11′
(g) log cosec 71° 10′   (h) log cosec 8° 9′   (i) log sec 11° 24′
(j) log sec 29° 3′   (k) log cot 40° 7′   (l) log cot 18° 29′

2) From the tables find the angle θ if:

(a) cosec θ is 1.352 7   (b) sec θ is 1.852   (c) cot θ is 0.491 7

3) Find the lengths of the sides marked x in Fig. 1.30. All the triangles are right angled.

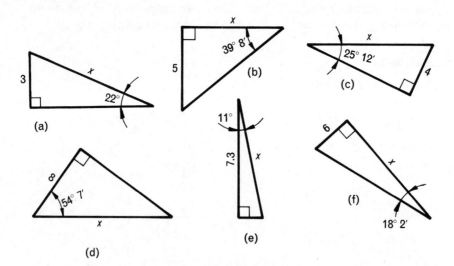

Fig. 1.30

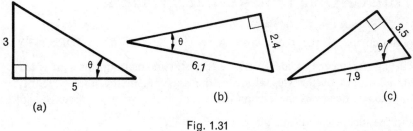

Fig. 1.31

4) By using the cosec, sec or cot find the angles marked θ in the triangles shown in Fig. 1.31. All the triangles are right angled.

5) Calculate the side of the triangle which is marked x in Fig. 1.32.

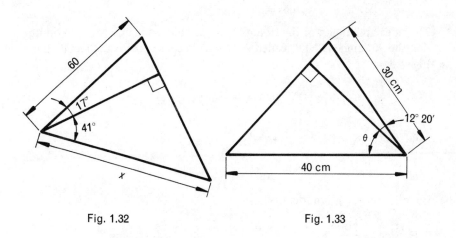

Fig. 1.32   Fig. 1.33

6) Calculate the angle θ in the triangle in Fig. 1.33.

7) The height of an isosceles triangle is 4.3 cm and each of the equal angles is 39°. Find the lengths of the equal sides.

8) Draw a triangle with sides 6 cm, 8 cm, and 10 cm long. Find the cosec, sec and cot of each of the acute angles. Hence find the angles from the tables and check these against your drawing.

9) The chord of a circle is 4.5 cm long and it subtends an angle of 71° at the centre. Calculate the radius of the circle.

10) 20 holes are equally spaced around the circumference of a circle. If the distance between the centres of two adjacent holes, measured along the chord, is 17 mm what is the diameter of the pitch circle?

TRIGONOMETRICAL IDENTITIES

A statement of the type $\cosec A \equiv \dfrac{1}{\sin A}$ is called an IDENTITY. The sign $\equiv$ means 'is identical to'. Any statement using this sign is true for all values of the variables, i.e. the angle A in the above identity. In practice, however, the $\equiv$ sign is often replaced by the $=$ (equals) sign. and the identity would be given as $\cosec A = \dfrac{1}{\sin A}$.

Many trigonometrical identities may be verified by the use of a right angled triangle.

EXAMPLE 12

To show that $\tan A = \dfrac{\sin A}{\cos A}$.

The sides and angles of the triangle may be labelled in any way providing that the 90° angle is NOT called A. We have chosen the usual labelling in Fig. 1.34.

Now $$\sin A = \frac{a}{b}$$

and $$\cos A = \frac{c}{b}$$

and $$\tan A = \frac{a}{c}$$

Hence from the given identity,

R.H. Side $= \dfrac{\sin A}{\cos A} = \dfrac{a/b}{c/b} = \dfrac{a.b}{b.c} = \dfrac{a}{c} = \tan A =$ L.H. Side

EXAMPLE 13

To show that $\cot A = \dfrac{\cos A}{\sin A}$.

Using again Fig. 1.34 we have:

$$\sin A = \frac{a}{b}$$

and $$\cos A = \frac{c}{b}$$

and $$\cot A = \frac{c}{a}$$

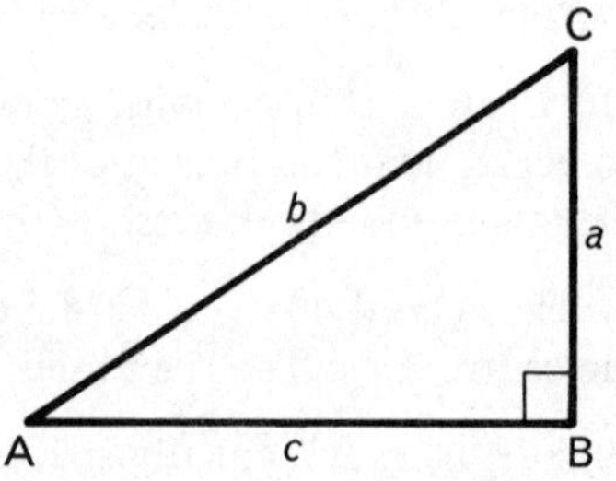

Fig. 1.34

Hence from the given identity,

R.H. Side $= \dfrac{\cos A}{\sin A} = \dfrac{c/b}{a/b} = \dfrac{c.b}{b.a} = \dfrac{c}{a} = \cot A =$ L.H. Side

Exercise 3

Prove the following trigonometrical identities using a right angled triangle:

1) $\tan A = \sin A.\sec A$

2) $\cot B = \cos B.\operatorname{cosec} B$

3) $\operatorname{cosec} C = \sec C.\cot C$

4) $\sec A = \operatorname{cosec} A.\tan A$

5) $\dfrac{\cot \theta}{\operatorname{cosec} \theta} = \cos \theta$

6) $\operatorname{cosec} B = \dfrac{1}{\cos B.\tan B}$

TRIGONOMETRICAL RATIOS BETWEEN 0° AND 360°

The sine, cosine and tangent of an angle between 0° and 90° have been previously defined. We now show how to deal with angles between 0° and 360°.

From Fig. 1.35, by definition,

$$\sin \theta = \frac{PM}{OP}$$

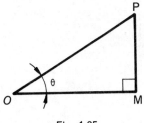

Fig. 1.35

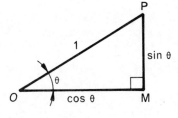

Fig. 1.36

If we make $OP = 1$ unit as shown in Fig. 1.36 then $\sin \theta = $ PM and $\cos \theta = $ OM.

In Fig. 1.37, the axes XOX' and YOY' have been drawn at right-angles to each other to form the four quadrants shown in the diagram. Drawing the axes in this way allows us to use the same sign convention as when drawing a graph.

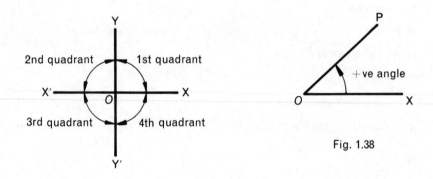

Fig. 1.37

Fig. 1.38

An angle, if positive, is measured in an anti-clockwise direction from the axis OX. It is formed by rotating a line, such as OP (Fig. 1.38) in an anti-clockwise direction.

Now if we draw a circle whose radius is OP we see that OP, as it rotates, forms the angle θ. If OP is made equal to 1 unit, then by drawing the right-angled triangle OPM (Fig. 1.39) the vertical height PM gives the value of $\sin \theta$ and the horizontal distance OM gives the value of $\cos \theta$.

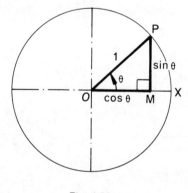

Fig. 1.39

The idea can be extended to angles greater than 90° as shown in Fig, 1.40.

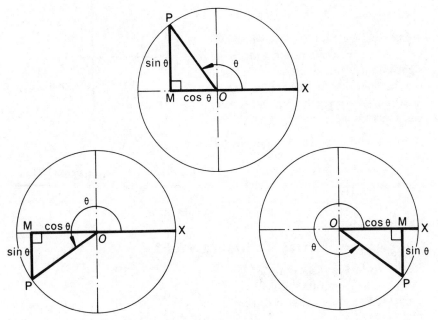

Fig. 1.40

We now make use of the sine convention used when drawing a graph. This means that when the height PM lies above the horizontal axis it is a positive length and when it lies below the horizontal axis it is a negative length. Hence when PM lies above the axis XOX' sin θ is positive and when it lies below the axis XOX' sin θ is negative.

Similarly, when the distance OM lies to the right of the origin O, it is regarded as being a positive distance; if it lies to the left of O it is regarded as being a negative distance. Hence when OM lies to the right of O, cos θ is positive and when it lies to the left of O, cos θ is negative.

This gives us the signs of the ratios sine and cosine in the four quadrants, see Fig. 1.41.

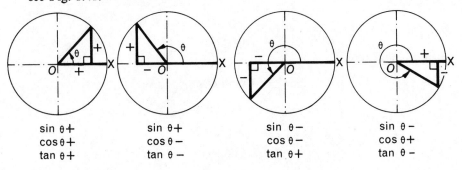

sin θ+	sin θ+	sin θ-	sin θ-
cos θ+	cos θ-	cos θ-	cos θ+
tan θ+	tan θ-	tan θ+	tan θ-

Fig. 1.41

We have previously shown that $\tan \theta \equiv \dfrac{\sin \theta}{\cos \theta}$.

Remembering that like signs, when divided, give a positive result and unlike signs give a negative result, we find that $\tan \theta$ is positive in the first and third quadrants and negative in the second and fourth quadrants.

All of the above results are summarised in Fig. 1.42.

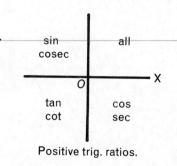

Positive trig. ratios.

Fig. 1.42

EXAMPLE 14

Find the values of sin 158°, cos 158° and tan 158°.

As shown in Fig. 1.43, sin 158° is given by the length PM. But from △ OPM, PM gives the sine of the angle POM,

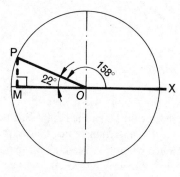

∴ $\sin 158° = \sin P\hat{O}M$

$= \sin (180° - 158°)$

$= \sin 22°$

$= 0.374\,6$

Fig. 1.43

Also from Fig. 1.43, OM gives the value of cos 158°: but this is a negative length since it lies to the left of origin O,

∴ $\cos 158° = -\cos P\hat{O}M$

$= -\cos (180° - 158°)$

$= -\cos 22°$

$= -0.927\,2$

In a similar way,

$\tan 158° = -\tan P\hat{O}M$

$= -\tan (180° - 158°)$

$= -\tan 22°$

$= -0.404\,0$

EXAMPLE 15

Find the values of sin 247°, cos 247° and tan 247°.

From Fig. 1.44,

$$\sin 247° = \text{MP} = -\sin \text{P}\hat{\text{O}}\text{M}$$
$$= -\sin (247° - 180°)$$
$$= -\sin 67°$$
$$= -0.920\,5$$
$$\cos 247° = O\text{M} = -\cos \text{P}\hat{\text{O}}\text{M}$$
$$= -\cos 67°$$
$$= -0.390\,7$$
$$\tan 247° = \frac{\text{MP}}{O\text{M}} = +\tan \text{P}\hat{\text{O}}\text{M}$$
$$= +\tan 67°$$
$$= 2.355\,9$$

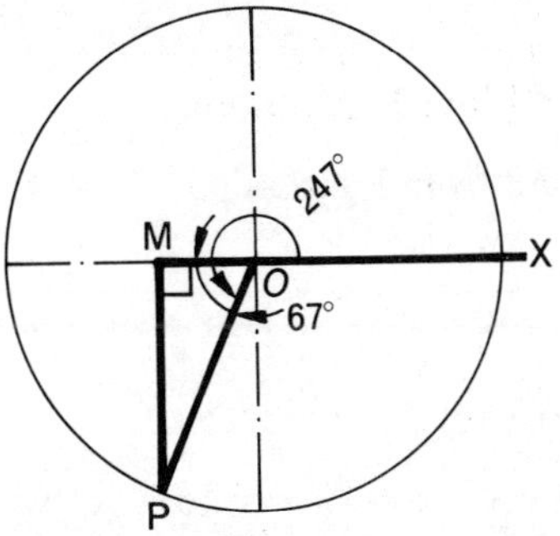

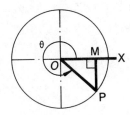

Fig. 1.44

EXAMPLE 16

Find all the angles between 0° and 360° whose:

(a)   sines are $-0.467\,6$,

(b)   cosines are $0.357\,2$,

(c)   cotangents are $-0.982\,7$.

a)   Let $\sin \theta = -0.467\,6$.

Since the sine is negative the angles θ lie in the 3rd and 4th quadrants. These are shown in Figs. 1.45 and 1.46.

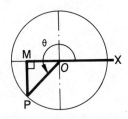

Fig. 1.45

Fig. 1.46

From right angled triangle OPM,

$$\sin \hat{POM} = MP = 0.467\,6$$

∴ $\hat{POM} = 27° 53'$ from tables.

From Fig. 1.45,

$$\theta = 180° + 27° 53'$$
$$= 207° 53'$$

and from Fig. 1.46,

$$\theta = 360° - 27° 53'$$
$$= 332° 7'$$

b)   Let $\cos \theta = 0.357\,2$.

Since the cosine is positive the angles θ lie in the 1st and 4th quadrants. These are shown in Figs. 1.47 and 1.48.

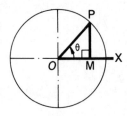   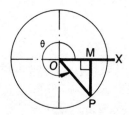

         Fig. 1.47   Fig. 1.48

From the right angle triangle OPM,

$$\cos \hat{POM} = OM = 0.357\,2$$

∴ $\hat{POM} = 69° 4'$ from tables.

From Fig. 1.47,

$$\theta = 69° 4'$$

and from Fig. 1.48,

$$\theta = 360° - 69° 4'$$
$$= 290° 56'$$

c)   Let $\cot \theta = -0.982\,7$.

Cotangents are treated in a similar manner to tangents, i.e. since the cotangent is negative the angles θ lie in the 2nd and 4th quadrants. These are shown in Figs. 1.49 and 1.50.

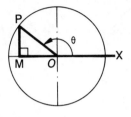   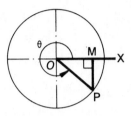

Fig. 1.49   Fig. 1.50

From the right angled triangle OPM,

$$\cot \hat{P O M} = \frac{OM}{MP} = 0.982\,7$$

$$\therefore \qquad \hat{P O M} = 45° \, 30' \quad \text{from tables.}$$

From Fig. 1.49,

$$\theta = 180° - 45° \, 30'$$

$$= 134° \, 30'$$

and from Fig. 1.50,

$$\theta = 360° - 45° \, 30'$$

$$= 314° \, 30'$$

The following tables may be used for angles in any quadrant:

Quadrant	Angle	$\sin \theta =$	$\cos \theta =$	$\tan \theta =$
first	0°—90°	$\sin \theta$	$\cos \theta$	$\tan \theta$
second	90°—180°	$\sin (180° - \theta)$	$-\cos (180° - \theta)$	$-\tan (180° - \theta)$
third	180°—270°	$-\sin (\theta - 180°)$	$-\cos (\theta - 180°)$	$\tan (\theta - 180°)$
fourth	270°—360°	$-\sin (360° - \theta)$	$\cos (360° - \theta)$	$-\tan (360° - \theta)$

Quadrant	Angle	$\operatorname{cosec} \theta =$	$\sec \theta =$	$\cot \theta =$
first	0°—90°	$\operatorname{cosec} \theta$	$\sec \theta$	$\cot \theta$
second	90°—180°	$\operatorname{cosec} (180 - \theta)$	$-\sec (180° - \theta)$	$-\cot (180° - \theta)$
third	180°—270°	$-\operatorname{cosec} (\theta - 180°)$	$-\sec (\theta - 180°)$	$\cot (\theta - 180°)$
fourth	270° - 360°	$-\operatorname{cosec} (360° - \theta)$	$\sec (360° - \theta)$	$-\cot (360° - \theta)$

Exercise 4

1) Write down the values of the sine, cosine and tangent of the following angles:

(a) 121°   (b) 178° 23'   (c) 102° 29'   (d) 211°
(e) 239° 17'   (f) 258° 28'   (g) 318° 27'   (h) 297° 17'

2) Find the values of cosec, sec and cot of the angles in Question 1.

3) Evaluate: 5 sin 142° − 3 tan 148° + 3 cos 230°.

4) Evaluate: sin A cos B − sin B cos A   given that sin A = $\frac{3}{5}$ and tan B = $\frac{4}{3}$. A and B are both acute angles.

5) An angle A is in the 2nd quadrant. If sin A = 3/5 find, without using tables, the value of cos A and tan A.

6) If sin θ = 0.143 2 find all the possible values of θ up to 360°.

7) If cos θ = −0.892 7 find all the possible values of θ up to 360°.

8) Find the angles in the first and second quadrants:

(a) whose sine is 0.713 7   (b) whose cosine is −0.481 3
(c) whose tangent is 0.947 6   (d) whose tangent is −1.764 2

9) Find the angles in the first and second quadrants:

(a) whose cosecant is 1.815 8   (b) whose secant is −1.881 7
(c) whose cotangent is −0.543 2

10) If sin A = $\dfrac{a \sin B}{b}$ find the values of A between 0° and 360° when a = 7.26 mm, b = 9.15 mm and B = 18° 29'.

11) If cos C = $\dfrac{(a^2+b^2-c^2)}{2ab}$ find the values of C between 0° and 360° given that a = 1.26 cm, b = 1.41 cm and c = 2.13 cm.

SINE, COSINE AND TANGENT CURVES

The Sine Curve

Using the tables to find values of angles between 0° and 90°, and also the methods previously explained for angles 90° to 360° we can draw up a table of values as shown. Intervals of 30° have been chosen to illustrate this but more angles may be taken to obtain a more accurate graph.

θ	0	30	60	90	120	150	180
$y = \sin\theta$	0	0.500	0.866	1.000	0.866	0.500	0

θ	210	240	270	300	330	360
$y = \sin\theta$	-0.500	-0.866	-1.000	-0.866	-0.500	0

These values are plotted as shown in Fig. 1.51.

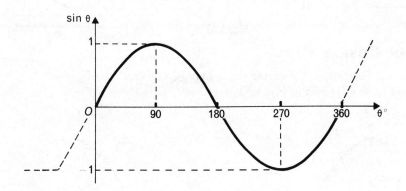

Fig. 1.51.

The graph of $y = \sin\theta$

The following features should be noted:

1) In the first quadrant as θ increases from 0° to 90°, sin θ increases from 0 to 1.

2) In the second quadrant as θ increases from 90° to 180°, sin θ decreases from 1 to 0.

3) In the third quadrant as θ increases from 180° to 270°, sin θ decreases from 0 to -1.

4) In the fourth quadrant as θ increases from 270° to 360°, sin θ increases from -1 to 0.

An alternative method of constructing the sine curve is shown in Fig. 1.52. When using the circle in connection with a sine curve the radius in different positions is called a 'rotating phasor'.

Constructing the sine curve by means of a rotating phasor.

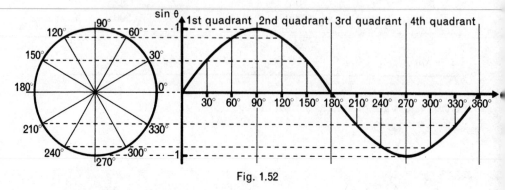

Fig. 1.52

The Cosine Curve

By a similar method the cosine curve may be constructed and it will be found to be as shown in Fig. 1.53.

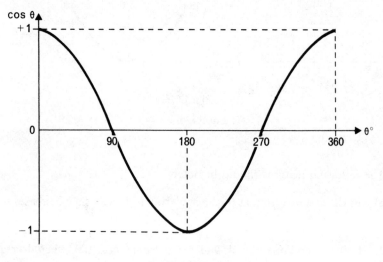

Fig. 1.53

The cosine curve

Note that:

1) In the first quadrant as θ increases from 0° to 90°, cos θ decreases from 1 to 0.

2) In the second quadrant as θ increases from 90° to 180°, cos θ decreases from 0 to -1.

3) In the third quadrant as θ increases from 180° to 270°, cos θ increases from -1 to 0.

4) In the fourth quadrant as θ increases from 270° to 360°, cos θ increases from 0 to 1.

The Tangent Curve

Using the methods as described for the sine and cosine curves the tangent curve may be obtained, and is shown in Fig. 1.54.

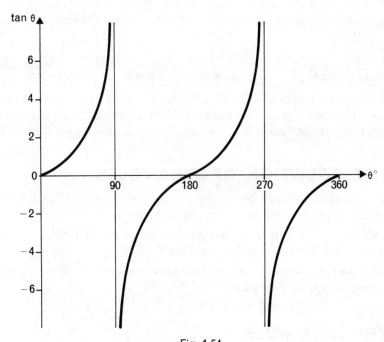

Fig. 1.54

The tangent curve

Note that:

1) In the first quadrant as θ increases from 0° to 90°, tan θ increases from 0 to infinity.

2) In the second quadrant as θ increases from 90° to 180°, tan θ increases from minus infinity to 0.

3) In the third quadrant as θ increases from 180° to 270° tan θ increases from 0 to infinity.

4) In the fourth quadrant as θ increases from 270° to 360° tan θ increases from minus infinity to 0.

Exercise 5

1) Draw the graphs of (a) $y = \sin \theta$ (b) $y = \cos \theta$ for values of θ between 0° and 360°. From the graphs find values of the sine and cosine of the angles:

(a) 38°	(b) 72°	(c) 142°	(d) 108°
(e) 200°	(f) 250°	(g) 305°	(h) 328°

2) Plot the graph of $y = 3 \sin x$ between 0° and 360°. From the graph read off the values of x for which $y = 1.50$ and find the value of y when $x = 250°$.

3) Draw the graphs of $3 \cos \theta$ for values of θ from 0° to 360°. Use the graph to find approximate values of the two angles for which $3 \cos \theta = 0.6$.

4) By projection from the circumference of a suitably marked off circle draw the graph of $4 \sin \theta$ for values of θ from 0° to 360°. Use the graph to find approximate values of the two angles for which $4 \sin \theta = 1.6$.

THE SOLUTION OF TRIANGLES

We now deal with triangles which are *not* right-angled. Every triangle consists of six elements — three angles and three sides.

If we are given any three of these six elements we can find the other three by using either the *Sine Rule* or the *Cosine Rule*.

When we have found the values of the three missing elements we are said to have 'solved the triangle'.

THE SINE RULE

The sine rule may be used when given:

a) One side and any two angles, or

b) Two sides and an angle opposite to one of the given sides. In this case two solutions may be found giving rise to what is called the 'ambiguous case', see Example 18.

Using the notation of Fig. 1.55 the *sine* rule states:

$$\frac{a}{\sin A} = \frac{b}{\sin B} = \frac{c}{\sin C}$$

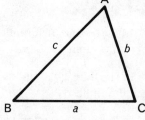

Fig. 1.55

EXAMPLE 17

Solve the triangle ABC given that
A = 42°, C = 72° and b = 61.8 mm.

The triangle should be drawn for
reference as shown in Fig. 1.56, but there
is no need to draw it to scale.

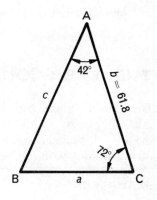

Fig. 1.56

Since $\angle A + \angle B + \angle C = 180°$
$\angle B = 180° - 42° - 72° = 66°$

The sine rule states:

$$\frac{a}{\sin A} = \frac{b}{\sin B}$$

$$\therefore \quad a = \frac{b \sin A}{\sin B}$$

$$= \frac{61.8 \times \sin 42°}{\sin 66°}$$

$$= 45.27 \text{ mm}$$

number	log
61.8	1.791 0
sin 42°	$\bar{1}$.825 5
	1.616 5
sin 66°	$\bar{1}$.960 7
45.27	1.655 8

Also,

$$\frac{c}{\sin C} = \frac{b}{\sin B}$$

$$\therefore \quad c = \frac{b \sin C}{\sin B}$$

$$= \frac{61.8 \times \sin 72°}{\sin 66°}$$

$$= 64.34 \text{ mm}$$

number	log
61.8	1.791 0
sin 72°	$\bar{1}$.978 2
	1.769 2
sin 66°	$\bar{1}$.960 7
64.34	1.808 5

The complete solution is:

$$\angle B = 66°, a = 45.27 \text{ mm}, c = 64.34 \text{ mm}$$

A rough check on sine rule calculations may be made by remembering
that in any triangle the longest side lies opposite the largest angle and the
shortest side lies opposite the smallest angle.

Thus in the previous example:

smallest angle $= 42° = A$; shortest side $= a = 45.27$ mm

largest angle $= 72° = C$; longest side $= c = 64.34$ mm

THE AMBIGUOUS CASE

There are two angles between 0° and 180° which have the same sine. For instance if sin A $= 0.5000$, then A can be either 30° or 150°. When using the sine rule to find an angle we must always examine the problem to see if there are two possible values for the angle.

EXAMPLE 18

In triangle ABC, $b = 93.23$ mm, $c = 85.61$ mm and $\angle C = 37°$. Solve the triangle.

Referring to Fig. 1.57 and using

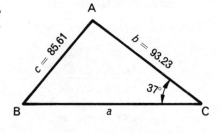

$$\frac{b}{\sin B} = \frac{c}{\sin C}$$

$$\therefore \quad \sin B = \frac{b \sin C}{c}$$

Fig. 1.57

$$= \frac{93.23 \times \sin 37°}{85.61}$$

$$= 0.6552$$

The angle B may be in either the first or second quadrants.

In the first quadrant, $\angle B = 40° 56'$ (Fig. 1.58)

In the second quadrant, $\angle B = 139° 4'$ (Fig. 1.59)

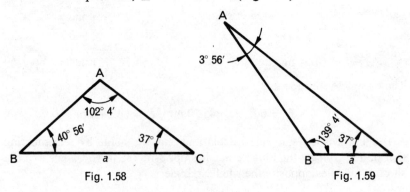

Fig. 1.58   Fig. 1.59

When $\angle B = 40° 56'$

$\angle A = 180° - 40° 56' - 37°$

$\qquad = 102° 4'$

Now

$$\frac{a}{\sin A} = \frac{c}{\sin C}$$

$\therefore \qquad a = \frac{c \sin A}{\sin C}$

$$= \frac{85.61 \sin 102° 4'}{\sin 37°}$$

$$= \frac{85.61 \sin 77° 56'}{\sin 37°}$$

$$= 139.0 \text{ mm}$$

When $\angle B = 139° 4'$

$\angle A = 180° - 139° 4' - 37°$

$\qquad = 3° 56'$

Now

$$\frac{a}{\sin A} = \frac{c}{\sin C}$$

$\therefore \qquad a = \frac{c \sin A}{\sin C}$

$$= \frac{85.61 \sin 3° 56'}{\sin 37°}$$

$$= 9.77 \text{ mm}$$

The ambiguous case may been seen clearly by constructing the given triangle geometrically as follows (Fig. 1.60).

Using a scale of one tenth full size draw AC = 9.323 cm and draw CX such that $A\hat{C}X = 37°$. Now with centre A and a radius 8.561 cm describe a circular arc to cut CX at B and B'.

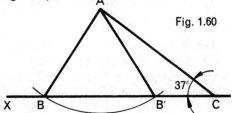

Fig. 1.60

Then ABC represents the triangle shown in Fig. 1.58 and AB'C represents the triangle shown in Fig. 1.59.

Use of the Sine Rule to Find the Diameter (D) of the Circumscribing Circle of a Triangle.

Using the notation of Fig. 1.61

$$\frac{a}{\sin A} = \frac{b}{\sin B} = \frac{c}{\sin C} = D$$

The rule is useful when we wish to find the pitch circle diameter of a ring of holes.

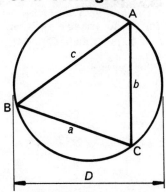

Fig. 1.61

EXAMPLE 19

In Fig. 1.62 three holes are positioned by the angle and dimensions shown. Find the pitch circle diameter.

We are given

$$\angle B = 41° \text{ and } b = 112.5 \text{ mm}$$

$$\therefore \quad D = \frac{b}{\sin B} = \frac{112.5}{\sin 41°}$$

$$= 171.5 \text{ mm}$$

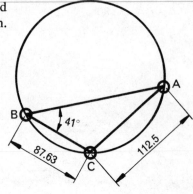

Fig. 1.62

THE COSINE RULE

The cosine rule is used in all cases where the sine rule cannot be used. These are when given:

a) two sides and the angle between them;

b) three sides.

Whenever possible the sine rule is used because it results in a calculation which is easier to perform. In solving a triangle it is sometimes necessary to start with the cosine rule and then having found one of the unknown elements to finish solving the triangle using the sine rule.

The cosine rule states:

either $\qquad a^2 = b^2 + c^2 - 2bc \cos A$

or $\qquad b^2 = a^2 + c^2 - 2ac \cos B$

or $\qquad c^2 = a^2 + b^2 - 2ab \cos C$

EXAMPLE 20

Solve the triangle ABC if $a = 70$ mm, $b = 40$ mm and $\angle C = 64°$.

Fig. 1.63

Referring to Fig. 1.63, to find the side c we use:

$$c^2 = a^2 + b^2 - 2ab \cos C$$

$$= 70^2 + 40^2 - 2 \times 70 \times 40 \times \cos 64°$$

$$= 4044$$

$$\therefore \qquad c = \sqrt{4044} = 63.6 \text{ mm}$$

We now use the sine rule to find $\angle A$:

$$\frac{a}{\sin A} = \frac{c}{\sin C}$$

$$\therefore \qquad \sin A = \frac{a \sin C}{c} = \frac{70 \times \sin 64°}{63.6}$$

$$\therefore \qquad A = 81° 36'$$

Hence $\qquad B = 180° - 81° 36' - 64° = 34° 24'$

EXAMPLE 21

Three holes A, B and C are located by the dimensions shown in Fig. 1.64. Calculate the centre distance between holes A and C.

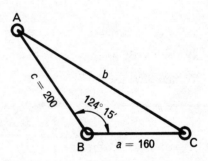

Fig. 1.64

Using the cosine rule:

$$b^2 = a^2 + c^2 - 2ac \cos B$$

$$= 160^2 + 200^2 - 2 \times 160 \times 200 \cos 124° 15'$$

Since

$$\cos 124° 15' = -\cos (180° - 124° 15')$$

$$= -\cos 55° 45' = -0.562\,8$$

then

$$b^2 = 160^2 + 200^2 - 2 \times 160 \times 200 \times (-0.562\,8)$$

$$= 160^2 + 200^2 + 2 \times 160 \times 200 \times 0.562\,8$$

$$\therefore \qquad b = 318.7 \text{ mm}$$

Hence the centre distance between holes A and C is 318.7 mm.

EXAMPLE 22

The diagram (Fig. 1.65), shows a slider-crank mechanism which is used in a feed mechanism. Calculate the crank angle θ:

a) when distance AC = 20.5 cm,

b) when distance AC = 11 cm.

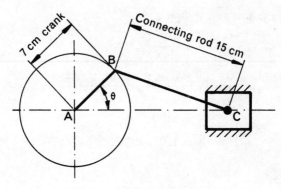

Fig. 1.65

a) In $\triangle ABC$ we have $a = 15$ cm, $b = 20.5$ cm and $c = 7$ cm.

To find θ we use

$$a^2 = b^2 + c^2 - 2bc\,(\cos\theta)$$

$$\therefore \qquad \cos\theta = \frac{b^2 + c^2 - a^2}{2bc}$$

$$= \frac{20.5^2 + 7^2 - 15^2}{2 \times 20.5 \times 7}$$

$$= 0.851\,1$$

$$\therefore \qquad \theta = 31° \, 40'$$

b) In $\triangle ABC$ we have $a = 15$ cm, $b = 11$ cm and $c = 7$ cm.
To find θ we use

$$a^2 = b^2 + c^2 - 2bc\,(\cos\theta)$$

$$\therefore \qquad \cos\theta = \frac{b^2 + c^2 - a^2}{2bc}$$

$$= \frac{11^2 + 7^2 - 15^2}{2 \times 11 \times 7}$$

$$= -0.357\,2$$

Because of the negative cosine, the angle θ must be in the second quadrant,

$$\therefore \qquad \cos(180° - \theta) = 0.357\,2$$

$$\therefore \qquad 180° - \theta = 69°\,4'$$

$$\therefore \qquad \theta = 110°\,56'$$

Exercise 6

1) The following are all exercises on the sine rule. Solve the following triangles ABC given:

(a)	$A = 75°$	$B = 34°$	$a = 10.2$ cm
(b)	$C = 61°$	$B = 71°$	$b = 91$ mm
(c)	$A = 19°$	$C = 105°$	$c = 11.1$ m
(d)	$A = 116°$	$C = 18°$	$a = 17$ cm
(e)	$A = 36°$	$B = 77°$	$b = 2.5$ m
(f)	$A = 49°\,11'$	$B = 67°\,17'$	$c = 11.22$ mm
(g)	$A = 17°\,15'$	$C = 27°\,7'$	$b = 22.15$ cm
(h)	$A = 77°\,3'$	$C = 21°\,3'$	$a = 9.793$ m
(i)	$B = 115°\,4'$	$C = 11°\,17'$	$c = 516.2$ mm
(j)	$a = 17$ cm	$b = 15$ cm	$B = 39°$
(k)	$a = 7$ m	$c = 11$ m	$C = 22°\,7'$
(l)	$b = 92$ mm	$c = 71$ mm	$C = 39°\,8'$
(m)	$b = 15.13$ cm	$c = 11.62$ cm	$B = 85°\,17'$
(n)	$a = 23$ cm	$c = 18.2$ cm	$A = 49°\,19'$
(o)	$a = 9.217$ cm	$b = 7.152$ cm	$A = 105°\,4'$

2) Solve the following triangles ABC using the cosine rule:

(a)	$a = 9$ cm	$b = 11$ cm	$C = 60°$
(b)	$b = 10$ cm	$c = 14$ cm	$A = 56°$
(c)	$a = 8.16$ m	$c = 7.14$ m	$B = 37°\,18'$
(d)	$a = 5$ m	$b = 8$ m	$c = 7$ m
(e)	$a = 312$ mm	$b = 527.3$ mm	$c = 700$ mm
(f)	$a = 7.912$ cm	$b = 4.318$ cm	$c = 11.08$ cm

3) Three holes lie on a pitch circle and their chordal distances are 41.82 mm, 61.37 mm and 58.29 mm. Find their pitch circle diameter.

4) In Fig. 1.66 find the angle BCA given that BC is parallel to AD.

5) In Fig. 1.67 find:

(a) the distance AB;
(b) the angle ACB.

6) Three holes are spaced as shown in Fig. 1.68. Calculate the centre distances from A to B and from A to C.

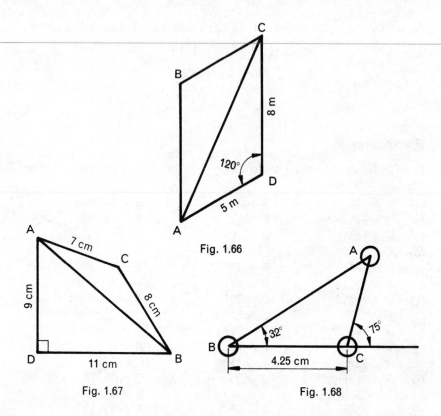

Fig. 1.66

Fig. 1.67

Fig. 1.68

7) Find the smallest angle in a triangle whose sides are 20, 25 and 30 m long.

8) In a slider-crank mechanism the connecting rod is 18 cm long and the crank is 5 cm long. At a particular instant the crank is 63° from inner dead centre position. Find the angle between the crank and the connecting rod.

9) Calculate the angle θ in Fig. 1.69. There are 12 castellations and they are equally spaced.

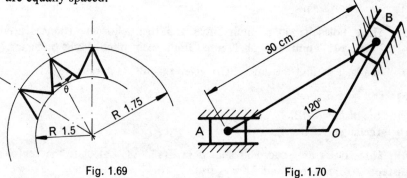

Fig. 1.69

Fig. 1.70

10) The mechanism shown in Fig. 1.70 consists of two pistons, which slide in cylinders whose axes are inclined to each other at 120°, and also a connecting rod 30 cm long. Find the distance AO when OB is 20 cm.

AREA OF A TRIANGLE

Three formulae are commonly used for finding the areas of triangles:
1) If given the base and the altitude (i.e. vertical height).
2) If given any two sides and the included angle.
3) If given the three sides.

Case 1) Given the base and the altitude.

In Fig. 1.71,

Area of triangle $= \frac{1}{2} \times$ base $\times$ altitude

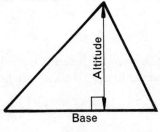

Fig. 1.71

EXAMPLE 23

Find the areas of the triangles shown in Fig. 1.72.

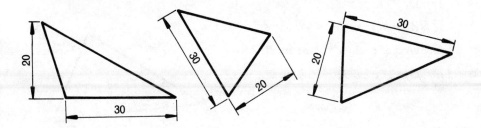

Fig. 1.72

In each case the 'base' is taken as the side of given length and the 'altitude' is measured perpendicular to this side.

Hence   Triangular area $= \frac{1}{2} \times$ base $\times$ altitude

$$= \frac{1}{2} \times 30 \times 20$$

$$= 300 \text{ mm}^2 \quad \text{in each case.}$$

EXAMPLE 24

A trapezium is shown in Fig. 1.73 in which AB is parallel to DC. Find its area.

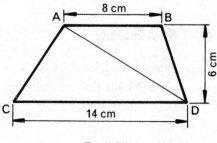

Fig. 1.73

If we join AD then the trapezium is divided into two triangles, the 'bases' and 'altitudes' of which are known.

Hence   area of trapezium = area of $\triangle$ABD + area of $\triangle$ADC

$$= \quad \tfrac{1}{2} \times 8 \times 6 \quad + \quad \tfrac{1}{2} \times 14 \times 6$$

$$= \quad\quad 24 \quad\quad + \quad\quad 42$$

$$= 66 \text{ cm}^2$$

Case 2)   If given any two sides and the included angle.

In Fig. 1.74,

Area of triangle = $\tfrac{1}{2}bc \sin A$

or   *area of triangle* = $\tfrac{1}{2}ac \sin B$

or   *area of triangle* = $\tfrac{1}{2}ab \sin C$

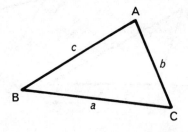

Fig. 1.74

EXAMPLE 25

Find the area of the triangle shown in Fig. 1.75.

Area = $\tfrac{1}{2} \times a \times c \times \sin B$

$= \tfrac{1}{2} \times 4 \times 3 \times \sin 30°$

$= 3 \text{ cm}^2$

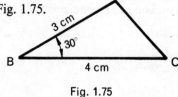

Fig. 1.75

EXAMPLE 26

Find the area of the triangle shown in Fig. 1.76.

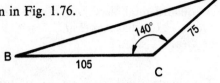

$\text{Area} = \tfrac{1}{2}ab \sin C$

$\qquad = \tfrac{1}{2} \times 105 \times 75 \times \sin 140°$

Fig. 1.76

We find the value of sin 140° by using the method explained previously as shown in Fig. 1.77, from which it may be seen that

$\qquad \sin 140° = \sin (180° - 140°)$

$\qquad\qquad\quad = \sin 40°$

$\therefore \qquad \text{Area} = \tfrac{1}{2} \times 105 \times 75 \times \sin 40°$

$\qquad\qquad\quad = 2531 \text{ mm}^2$

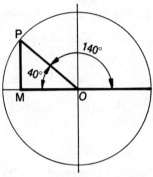

Fig. 1.77

Case 3) If given the three sides.

In Fig. 1.78,

Area of triangle $= \sqrt{s(s-a)(s-b)(s-c)}$

where $\qquad s = \dfrac{a+b+c}{2}$

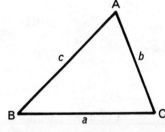

Fig. 1.78

EXAMPLE 27

A triangle has sides of lengths 3 cm, 5 cm and 6 cm. What is its area?

Since we are given the lengths of 3 sides we use

$$\text{area} = \sqrt{s(s-a)(s-b)(s-c)}$$

Now, $\qquad\qquad s = \dfrac{3+5+6}{2} = 7$

$\therefore \qquad\qquad \text{area} = \sqrt{7 \times (7-3) \times (7-5) \times (7-6)}$

$\qquad\qquad\qquad = \sqrt{7 \times 4 \times 2 \times 1}$

$\qquad\qquad\qquad = \sqrt{56} = 7.483 \text{ cm}^2$

EXAMPLE 28

A quadrilateral has the dimensions shown in the diagram (Fig. 1.79). Find its area.

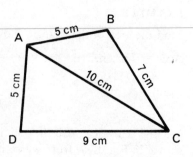

The quadrilateral is made up of the triangles ABC and ACD.

To find the area of $\triangle$ ABC.

Fig. 1.79

$$s = \frac{5+7+10}{2} = 11$$

$\therefore$ area of $\triangle$ ABC $= \sqrt{s(s-a)(s-b)(s-c)}$

$$= \sqrt{11 \times (11-5)(11-7)(11-10)}$$

$$= \sqrt{11 \times 6 \times 4 \times 1}$$

$$= \sqrt{264} = 16.25 \text{ cm}^2$$

To find the area of $\triangle$ ACD,

$$s = \frac{5+9+10}{2} = 12$$

$\therefore$ area of $\triangle$ ACD $= \sqrt{s(s-a)(s-b)(s-c)}$

$$= \sqrt{12(12-5)(12-9)(12-10)}$$

$$= \sqrt{12 \times 7 \times 3 \times 2}$$

$$= \sqrt{504} = 22.45 \text{ cm}^2$$

$\therefore$ area of quadrilateral $=$ area of $\triangle$ ABC $+$ area of $\triangle$ ACD

$$= 16.25 + 22.45 = 38.70 \text{ cm}^2$$

Exercise 7

1) Find the area of a triangle whose base is 7.5 cm and whose altitude is 5.9 cm.

2) Find the area of an isosceles triangle whose equal sides are 8.2 cm and whose base is 9.5 cm.

3) A plate in the shape of an equilateral triangle has a mass of 12.25 kg. If the material has a mass of 3.7 kg/m² find the dimensions of the plate in mm.

4) Obtain the area of a triangle whose sides are 39.3 cm and 41.5 cm if the angle between them is 41° 30′.

5) Find the area of the template shown in Fig. 1.80.

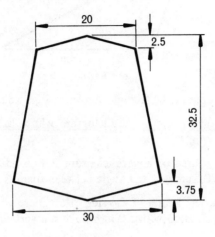

Fig. 1.80

6) Calculate the area of a triangle ABC if:
(a) $a = 4$ cm, $b = 5$ cm and $\angle C = 49°$,
(b) $a = 3$ m, $c = 6$ m and $\angle B = 63° 44′$.

7) A triangle has sides 4 cm, 7 cm and 9 cm long. What is its area?

8) A triangle has sides 37 mm, 52 mm and 63 mm long. What is its area in cm²?

9) Find the areas of the quadrilaterals shown in Fig. 1.81.

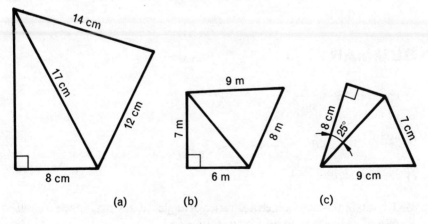

Fig. 1.81

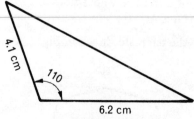

Fig. 1.82

10) Find the area of the triangle shown in Fig. 1.82.

11) What is the area of a parallelogram whose base is 7 cm long and whose vertical height is 4 cm?

12) Obtain the area of a parallelogram if two adjacent sides measure 11.25 cm and 10.5 cm and the angle between them is 49°.

13) Determine the length of the side of a square whose area is equal to that of a parallelogram with a 3 m base and a vertical height of 1.5 m.

14) Find the area of a trapezium whose parallel sides are 75 mm and 82 mm long respectively and whose vertical height is 39 mm.

15) Find the area of a regular hexagon,
(a) which is 4 cm wide across flats,
(b) which has sides 5 cm long.

16) Find the area of a regular octagon,
(a) which is 2 mm wide across flats,
(b) which has sides 2 mm long.

17) The parallel sides of a trapezium are 12 cm and 16 cm long. If its area is 220 cm² what is its altitude?

SUMMARY

a) $\operatorname{cosec(ant)} \theta = \dfrac{1}{\sin \theta}$.

b) $\operatorname{sec(ant)} = \dfrac{1}{\cos \theta}$.

c) $\operatorname{cotan(gent)} \theta = \dfrac{1}{\tan \theta}$.

d) To obtain the trigonometrical ratios of angles of any magnitude a circle of unit radius is used as shown in Fig. 1.83.

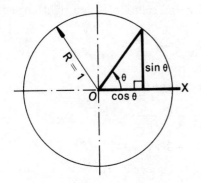

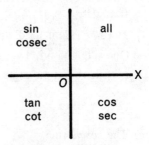

Trigonometrical ratios
which are positive in each
of the four quadrants.

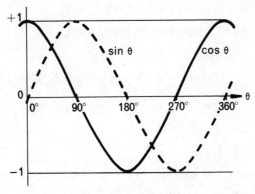

Curves of sin θ and cos θ

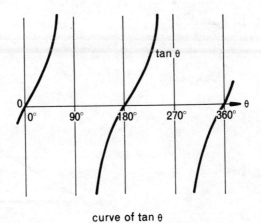

curve of tan θ

Fig. 1.83

e) The sine and cosine rules are used in the solution of non-right angled triangles.

f) The *sine* rule is used when given:

One side and any two angles, or

two sides and an angle opposite one of the sides.

$$\frac{a}{\sin A} = \frac{b}{\sin B} = \frac{c}{\sin C}$$

g) The *cosine* rule, $a^2 = b^2+c^2-2bc \cos A$

or $b^2 = a^2+c^2-2ac \cos B$

or $c^2 = a^2+b^2-2ab \cos C$,   used when given:

Two sides and the angle between them, or three sides.

h) The *area* of a triangle may be found using:

either   Area $= \frac{1}{2} \times$ base $\times$ altitude

or   Area $= \frac{1}{2}ab.\sin C = \frac{1}{2}bc.\sin A = \frac{1}{2}ac.\sin B$

or   Area $= \sqrt{s(s-a)(s-b)(s-c)}$ where $s = \dfrac{a+b+c}{2}$

Self Test 1

In each case state which answers are correct:

1) cosec C is:

a $\dfrac{1}{\cos C}$ **b** $\dfrac{1}{\tan C}$ **c** $\dfrac{1}{\sin C}$ **d** $\dfrac{1}{\sec C}$

2) In Fig. 1.84, sec A is:

a $\dfrac{b}{a}$ **b** $\dfrac{b}{c}$ **c** $\dfrac{a}{c}$ **d** $\dfrac{c}{a}$

3) In Fig. 1.84, cosec C is:

a $\dfrac{b}{a}$ **b** $\dfrac{c}{b}$

c $\dfrac{b}{c}$ **d** $\dfrac{a}{b}$

4) In Fig. 1.84, cot A is:

a $\dfrac{c}{a}$ **b** $\dfrac{a}{c}$

c $\dfrac{a}{b}$ **d** $\dfrac{b}{c}$

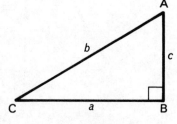

Fig. 1.84

5) In Fig. 1.84 the ratio $\dfrac{a}{c}$ is:

    a sin C   **b** cos A   **c** cot C   **d** tan A

6) The sum of the interior angles of a triangle is:

    a 90°   **b** 180°   **c** 270°   **d** 360°

7) In Fig. 1.85 angle A may be found using:

    a $\dfrac{a}{\sin A} = \dfrac{c}{\sin C}$ **b** $\dfrac{b}{\sin B} = \dfrac{c}{\sin C}$

    c $\dfrac{a}{\sin A} = \dfrac{b}{\sin B}$ **d** the sine rule

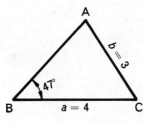

Fig. 1.85

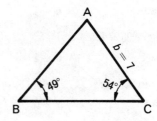

Fig. 1.86

8) In Fig. 1.86 side c may be found using:

    a $\dfrac{a}{\sin A} = \dfrac{b}{\sin B}$ **b** $\dfrac{b}{\sin B} = \dfrac{c}{\sin C}$

    c $\dfrac{c}{\sin C} = \dfrac{a}{\sin A}$ **d** the cosine rule

9) In Fig. 1.87 the side b may be found using:

    a $\dfrac{a}{\sin A} = \dfrac{c}{\sin C}$ **b** $\dfrac{a}{\sin A} = \dfrac{b}{\sin B}$

    c $b^2 = a^2 + c^2 - 2ac.\cos A$ **d** $\frac{1}{2}ac.\sin B$

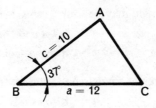

Fig. 1.87

10)   The cosine formula may be used for solving a triangle when given:
a any three sides.
b any two sides and the angle opposite one of them.
c one side and any two angles.
d two sides and the angle between them

11)   sin 127° is:

    a 1.798 6   **b** −1.798 6   **c** 0.798 6   **d** −0.798 6

12)   cos 154° is:

    a 0.898 8   **b** −0.898 8   **c** 1.101 2   **d** −1.101 2

13)   tan 180° is:

    a 0.5   **b** −1   **c** 1   **d** 0

14)   cos 270° is:

    a 1   **b** −1   **c** 0   **d** −0.342 0

15)   sin 90° is:

    a 0   **b** −0   **c** 1   **d** −1

16)   The angle whose sine is 0.5 is:

    a 30°   **b** 130°   **c** 150°   **d** 210°

17)   The angle whose cosine is −0.5 is:

    a 60°   **b** 120°   **c** 240°   **d** 300°

18)   The angle whose tangent is negative occurs in the

    a 1st quadrant   **b** 2nd quadrant **c** 3rd quadrant **d** 4th quadrant

19)   The angle whose tangent is zero is:

    a 0°   **b** 90°   **c** 180°   **d** 270°   **e** 360°

20)   The angle whose tangent is −1 is:

    a 45°   **b** 135°   **c** 225°   **d** 315°

21)   The area of a triangle may be found using the formula:

    a base × altitude   **b** $\frac{1}{2}$ × base × altitude
    c $\frac{1}{3}$ × base × altitude   **d** (base)² × altitude

22)   In Fig. 1.88, the area of triangle ABC may be found directly by using:

    a $\frac{1}{2}$ × c × altitude   **b** $\frac{1}{2}ac$.sin B
    c $\sqrt{s(s-a)(s-b)(s-c)}$ **d** $\frac{1}{2}ab$.sin C

23) In the formula $\sqrt{s(s-a)(s-b)(s-c)}$ the letter s represents:

a the perimeter $b\ a+b+c$

c the semi-perimeter $d\ \frac{1}{2}(a+b+c)$

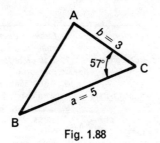

Fig. 1.88

2. MENSURATION

On reaching the end of this chapter you should be able to:-

1. *Calculate the volume and surface area of pyramids, cones and spheres.*
2. *Define frustum.*
3. *Calculate the surface area and volume of frusta of cones and pyramids.*
4. *Identify the component basic shapes of composite areas and volumes (i.e. rectangles, triangles, circles, prisms, pyramids, cones, cylinders and spheres).*
5. *Calculate the total area or volume of a composite figure.*
6. *Draw a diagram from given data to a suitable scale and use a planimeter to measure its area.*
7. *Calculate the area of irregular shapes using the mid-ordinate and trapezoidal rules.*
8. *Apply Simpson's rule to the determination of area using* 2n *strips* ($n > 3$).
9. *Calculate, given incremental cross-sectional areas, the volume of a solid using Simpson's rule.*
10. *State numerical results of any calculation in this section to an accuracy consistent with the data used.*

UNITS OF LENGTH

The standard measurement of length is the metre (abbreviation: m). This is split up into smaller units as follows:

$$\begin{aligned} 1 \text{ metre (m)} &= 10 \text{ decimetres (dm)} \\ &= 100 \text{ centimetres (cm)} \\ &= 1000 \text{ millimetres (mm)} \end{aligned}$$

For large measurements the kilometre is used and:

$$1 \text{ kilometre (km)} = 1000 \text{ metres (m)}$$

UNITS OF AREA

The area of a figure is measured by finding how many square units it contains. 1 square metre is the area inside a square which has a side of 1 metre (Fig. 2.1).

Similarly 1 square centimetre is the area inside a square whose side is 1 cm, and 1 square millimetre is the area inside a square whose side is 1 mm.

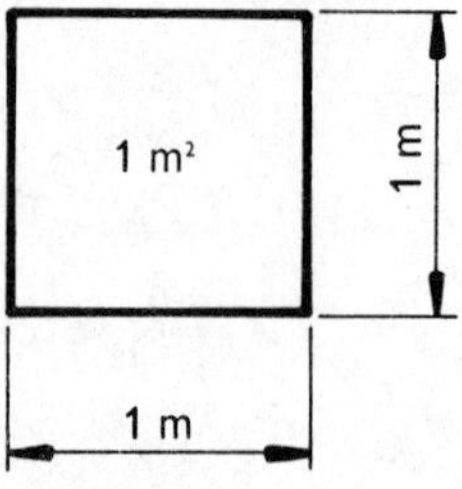

Fig. 2.1

The standard abbreviations for units of area are:

$$\text{square metres} = \text{m}^2$$
$$\text{square centimetres} = \text{cm}^2$$
$$\text{square millimetres} = \text{mm}^2$$

Conversions of square units:

$$1 \text{ m}^2 = (100 \text{ cm})^2 = (100 \times 100) \text{ cm}^2 = 10^4 \text{ cm}^2$$
$$1 \text{ m}^2 = (1000 \text{ mm})^2 = (1000 \times 1000) \text{ mm}^2 = 10^6 \text{ mm}^2$$
$$1 \text{ cm}^2 = (10 \text{ mm})^2 = (10 \times 10) \text{ mm}^2 = 10^2 \text{ mm}^2$$

EXAMPLE 1

A room in a factory is 11 m long and 7 m wide. It is to be covered with tiles, the area of each tile being 14 000 mm². How many tiles are needed?

$$\text{Area of room} = 11 \times 7 = 77 \text{ m}^2$$
$$= 77 \times 10^6 \text{ mm}^2$$

and

$$\text{area of tile} = 14\,000 = 14 \times 10^3 \text{ mm}^2$$

∴

$$\text{number of tiles required} = \frac{\text{area of room}}{\text{area of each tile}}$$

$$= \frac{77 \times 10^6}{14 \times 10^3} = 5.5 \times 10^3 = 5\,500$$

Exercise 8

1) A sheet of steel 2 m long by 1 m wide is cut up into 400 strips, each strip being 250 mm long. What is the width of each strip?

2) A workshop floor measures 8 m by 6 m. A roll of vinyl floor covering 1500 mm wide and 30 m long is laid. How many square metres of floor are left uncovered?

3) Blanks of area 20 cm² are being pressed out on a flypress from a sheet of aluminium 1000 mm long and 500 mm wide. Find the area of unused aluminium if 180 blanks are obtained from the sheet.

4) How many 30 cm square insulation tiles are needed to cover a ceiling 12 m long by 4.5 m wide, allowing for a 5% wastage?

UNITS OF VOLUME

The volume of an object is measured by finding how many cubic units it contains.

1 cubic metre (m³) is the volume of a cube whose side is 1 m (Fig. 2.2). Similarly 1 cubic centimetre (cm³) is the volume of a cube whose side is 1 cm.

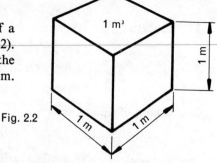

Fig. 2.2

Conversions of cubic units are:

$$1 \text{ m}^3 = (100 \text{ cm})^3 = (100 \times 100 \times 100) \text{ cm}^3 = 10^6 \text{ cm}^3$$

$$1 \text{ m}^3 = (1000 \text{ mm})^3 = (1000 \times 1000 \times 1000) \text{ mm}^3 = 10^9 \text{ mm}^3$$

$$1 \text{ cm}^3 = (10 \text{ mm})^3 = (10 \times 10 \times 10) \text{ mm}^3 = 10^3 \text{ mm}^3$$

EXAMPLE 2

How many cubes of side 20 mm are contained by a rectangular copper bar 3 m long, 10 cm wide and 4 cm thick?

$$\text{Volume of bar} = (3 \times 10^3) \times (10 \times 10) \times (4 \times 10) = 12 \times 10^6 \text{ mm}^3$$

$$\text{and} \quad \text{volume of cube} = 20 \times 20 \times 20 = 8 \times 10^3 \text{ mm}^3$$

$$\therefore \quad \text{number of cubes} = \frac{\text{volume of bar}}{\text{volume of cube}}$$

$$= \frac{12 \times 10^6}{8 \times 10^3} = 1.5 \times 10^3 = 1500$$

CAPACITY

The capacity of a container is the volume that it will contain. It is often measured in the same units as volume, that is cubic metres, cubic centimetres or cubic metres.

Sometimes however, as in the case of liquid measure the litre (abbreviation ℓ) unit is used. The litre is not a precise measurement because 1 litre = 1000.028 cm³. However, for most practical problems the litre can be assumed to be 1000 cm³.

Small capacities are often measured in millilitres (mℓ) and:

$$1000 \text{ millilitres (m}\ell) = 1 \text{ litre } (\ell)$$

but there are 1000 cm³ in 1 litre

$$\therefore \qquad\qquad\qquad 1 \text{ m}\ell = 1 \text{ cm}^3$$

EXAMPLE 3

A rectangular tank has inside measurements of 3 m by 2 m by 1.5 m. How many litres of liquid will it hold?

Since 1 litre is 1000 cm³ it is better to calculate the volume (or capacity) in cm³. Thus:

$$\text{Capacity} = 300 \times 200 \times 150 = 9\,000\,000 \text{ cm}^3$$

$$= \frac{9\,000\,000}{1000} = 9000 \text{ litres}$$

EXAMPLE 4

How many charges of liquid fuel, each of 3 millilitres may be obtained from a rectangular tank 25 cm by 12 cm by 7 cm?

$$\text{Volume of tank} = 25 \times 12 \times 7 = 2100 \text{ cm}^3$$

Now since there are 1000 cm³ to 1 litre, then 1 cm³ = 1 mℓ hence,

$$\text{number of 3 m}\ell \text{ charges} = \frac{2100}{3} = 700$$

DENSITY

The density of a substance is the mass per unit volume. Densities are usually measured in kilogrammes per cubic metre (kg/m³) or in grammes per cubic centimetre (g/cm³). The mass of an object may be found using the formula:

$$\text{Mass} = \text{density} \times \text{volume}$$

The table below gives the densities of various common substances:

Substance	Density (kg/m³)	Substance	Density (kg/m³)
Alcohol	790	Gravel	1800
Aluminium	2700	Ice	900
Asbestos	2800	Iron	7900
Brick, common	1800	Kerosene	800
Cement	3100	Lead	11 400
Coal, bituminous	1300	Masonry	2400
Concrete	2200	Petroleum oil	820
Copper	8900	Salt, common	2100
Gasoline	700	Sand, dry	1600
Glass	2600	Silver	10 500
Gold	19 300	Water, fresh	1000

EXAMPLE 5

Find the mass of a block of copper 5 cm by 6 cm by 8 cm, if the density of copper is 9 g/cm³.

$$\text{Volume of copper block} = 5 \times 6 \times 8 = 240 \text{ cm}^3$$

∴
$$\text{Mass of block} = \text{density} \times \text{volume}$$

$$= 9 \times 240$$

$$= 2160 \text{ g}$$

$$= 2.16 \text{ kg}$$

THE FLOW OF WATER

When a tank or container is being filled the time taken to fill the tank depends upon the quantity of water entering the tank in unit time. This is called the rate of flow and is often stated in cubic centimetres per second (cm³/s) or cubic metres per minute (m³/min).

In the case of water flowing through a pipe the rate of flow depends on the velocity or speed of flow as well as the area of the pipe cross-section. The relationship is:

$$\text{rate of flow} = \text{area of pipe cross-section} \times \text{velocity of flow}$$

EXAMPLE 6

A tank which contains 250 m³ of water when full is to be filled through a pipe which delivers water at a rate of 2 m³/min. How long does it take to fill the tank?

$$\text{Time taken} = \frac{\text{volume required to fill tank}}{\text{rate of flow}} = \frac{250}{2} = 125 \text{ min}$$

EXAMPLE 7

Water is flowing through a pipe whose area of cross-section is 4400 mm² at a speed of 2 m/s. Calculate the discharge from the pipe in:

a) cubic metres per second,
b) litres per minute.

a) The discharge is another name for rate of flow and since
$$1 \text{ m}^2 = 1\,000\,000 \text{ mm}^2$$

then $\text{discharge} = \text{area of pipe cross-section} \times \text{velocity of flow}$

$$= \frac{4400}{1\,000\,000} \times 2 = 0.088 \text{ m}^3/\text{s}$$

b) Since 1 m³ = 1 000 000 cm³ and 1 litre = 1000 cm³

then   1 m³ = 1000 litres

∴   discharge = 0.008 8 m³/s

   = 0.008 8 × 1000 ℓ/s

   = 8.8 ℓ/s

   = 8.8 × 60 = 528 ℓ/min

Exercise 9

1) A block of aluminium 2 m long, 200 mm wide and 50 mm deep is sawn up into 1280 cubes. Neglecting the width of the saw cuts find the length of the side of each cube.

2) An ingot containing 2 m³ of molten metal is made into strip 50 cm wide and 12 mm thick. What is the length of the strip?

3) A rectangular block of concrete measuring 45 cm by 25 cm by 15 cm has a mass of 40 kg. What is the density of the concrete?

4) 1 millilitre of a liquid has a mass of 0.8 gm. Find the mass in kilogrammes of 500 litres of this liquid.

5) The mass of a piece of silver is 200 g. If the density of silver is 10 g/cm³ what is the volume of the silver?

6) A domestic oil tank measures 1.5 m by 1 m by 1 m. Find its capacity in litres.

7) If the density of oil is 0.8 g/cm³ what is the mass of the oil in Question 6 if the tank is full?

8) Water is poured into a container at a rate of 300 cm³/s. How long does it take for the container to hold 150 ℓ?

9) Water is flowing through a pipe whose cross-sectional area is 200 cm² at a speed of 3 m/s. Find the discharge from the pipe:

(a) in cubic metres per second,
(b) in litres per minute.

10) Water flows along a channel at a velocity of 3 m/s. Find the cross-sectional area of the channel if the rate of flow is 180 000 ℓ/min.

11) Water is poured into a reservoir 15 m long and 10 m wide at the rate of 3000 litres per minute. At what rate does the water level rise in cm/min?

AREAS AND PERIMETERS

Rectangle

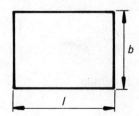

$$\text{Area} = l \times b$$
$$\text{Perimeter} = 2l + 2b$$

EXAMPLE 8

Find the area of the section shown in Fig. 2.3.

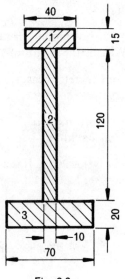

Fig. 2.3

The section can be split up into three rectangles as shown. The total area can be found by calculating the areas of the three rectangles separately and then adding these together. Thus,

$$\text{area of rectangle 1} = 15 \times 40 \ = 600 \text{ mm}^2$$

$$\text{area of rectangle 2} = 10 \times 120 = 1200 \text{ mm}^2$$

$$\text{area of rectangle 3} = 20 \times 70 \ = 1400 \text{ mm}^2$$

$$\text{Total area of section} = 600 + 1200 + 1400$$

$$= 3200 \text{ mm}^2$$

Parallelogram

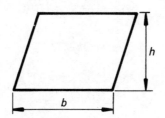

$$\text{Area} = b \times h$$

EXAMPLE 9

Find the area of the parallelogram shown in Fig. 2.4. The first step is to find the vertical height h.

In $\triangle BCE$,

$$h = 3 \times \sin 60° = 3 \times 0.866 = 2.598$$

Area of parallelogram = base × vertical height

$$= 5 \times 2.598 = 12.99 \text{ cm}^2$$

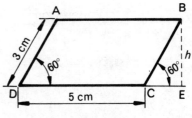

Fig. 2.4

Triangle

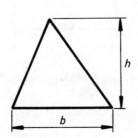

$$\text{Area} = \tfrac{1}{2} \times b \times h$$

(*see also* Chapter 1)

EXAMPLE 10

Find the area of a regular octagon (8-sided figure) which is 50 mm across flats (Fig. 2.5).

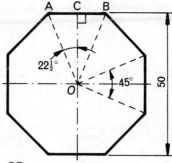

The angle subtended at the centre by a side of the octagon $= \dfrac{360°}{8} = 45°$

Now triangle AOB is isosceles, since OA = OB.

$$\therefore \qquad \angle AOC = \frac{45°}{2} = 22° \, 30'$$

Fig. 2.5

But $\qquad\qquad$ $OC = \dfrac{50}{2} = 25$ mm

Also $\qquad\qquad$ $\dfrac{AC}{OC} = \tan 22° \, 30'$

$\therefore$ $\qquad\qquad$ $AC = OC \times \tan 22° \, 30' = 25 \times 0.414 \, 2$

$\qquad\qquad\qquad\quad = 10.35$ mm

$\quad$ Area of $\triangle AOB = AC \times OC = 10.35 \times 25$

$\qquad\qquad\qquad\qquad = 258.8$ mm^2

$\quad$ Area of octagon $= 258.8 \times 8 = 2070$ mm^2

Trapezium

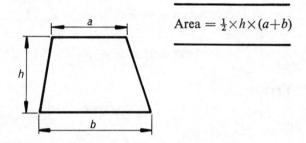

Area $= \frac{1}{2} \times h \times (a+b)$

EXAMPLE 11

The cross-section of a lathe slide is a trapezium with the dimensions shown. Find the area of the cross-section (Fig. 2.6).

Area $= \frac{1}{2} \times h \times (a+b)$

$\quad = \frac{1}{2} \times 40 \times (30+50)$

$\quad = \frac{1}{2} \times 40 \times 80$

$\quad = 1600$ mm^2

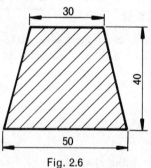

Fig. 2.6

Circle

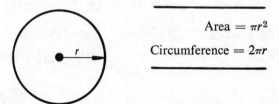

Area $= \pi r^2$

Circumference $= 2\pi r$

EXAMPLE 12

A hollow shaft has an outside diameter of 3.25 cm and an inside diameter of 2.5 cm. Calculate the cross-sectional area of the shaft (Fig. 2.7).

Area of cross-section = area of outside circle − area of inside circle

$$= \pi \times 1.625^2 - \pi \times 1.25^2$$

$$= \pi(1.625^2 - 1.25^2)$$

$$= 3.142 \times (2.640 - 1.563)$$

$$= 3.142 \times 1.077$$

$$= 3.38 \text{ cm}^2$$

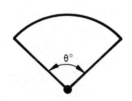

R 1.25 cm

R 1.625 cm

Fig. 2.7

Sector of a Circle

$$\text{Area} = \pi r^2 \times \frac{\theta^°}{360}$$

$$\text{Length of arc} = 2\pi r \times \frac{\theta^°}{360}$$

EXAMPLE 13

Calculate (**a**) the length of arc of a circle whose radius is 8 m and which subtends an angle of 56° at the centre, and (**b**) the area of the sector so formed.

a) $\text{Length of arc } = 2\pi r \times \dfrac{\theta^°}{360} = 2 \times \pi \times 8 \times \dfrac{56}{360}$

$$= 7.82 \text{ m}$$

b) $\text{Area of sector} = \pi r^2 \times \dfrac{\theta^°}{360} = \pi \times 8^2 \times \dfrac{56}{360}$

$$= 31.28 \text{ m}^2$$

EXAMPLE 14

In a circle of radius 40 mm a chord is drawn which subtends an angle of 120° at the centre. What is the area of the minor segment?

In Fig. 2.8 let A = area of sector of MCNO.

Then

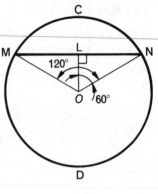

$$A = \frac{\pi r^2 \theta°}{360} = \frac{\pi \times 40^2 \times 120}{360}$$

$$= 1676 \text{ mm}^2$$

In the $\triangle MON$, $MO = NO = 40$ mm and the included angle $MON = 120°$.

Hence $\angle LON = 60°$

$$\frac{OL}{ON} = \cos 60°$$

Fig. 2.8

∴ $OL = ON \times \cos 60° = 40 \times 0.500\,0 = 20$ mm

$$\frac{LN}{ON} = \sin 60°$$

∴ $LN = ON \times \sin 60° = 40 \times 0.866\,0 = 34.64$ mm

$$\text{Area of } \triangle MON = \tfrac{1}{2} \times OL \times MN = \tfrac{1}{2} \times 20 \times 69.28$$

$$= 692.8 \text{ mm}^2$$

Area of minor segment MCNL $= 1676 - 692.8$

$$= 983.2 \text{ mm}^2$$

Exercise 10

1) The area of a metal plate is 220 mm². If its width is 25 mm, find its length.

2) A sheet metal plate has a length of 147.5 mm and a width of 86.5 mm. Find its area in m².

3) Find the areas of the sections shown in Fig. 2.9.

4) What is the area of a parallelogram whose base is 7 cm long and whose vertical height is 4 cm?

5) Obtain the area of a parallelogram if two adjacent sides measure 11.25 cm and 10.5 cm and the angle between them is 49°.

6) Determine the length of the side of a square whose area is equal to that of a parallelogram with a 3 m base and a vertical height of 1.5 m.

7) Find the area of a trapezium whose parallel sides are 75 mm and 82 mm long respectively and whose vertical height is 39 mm.

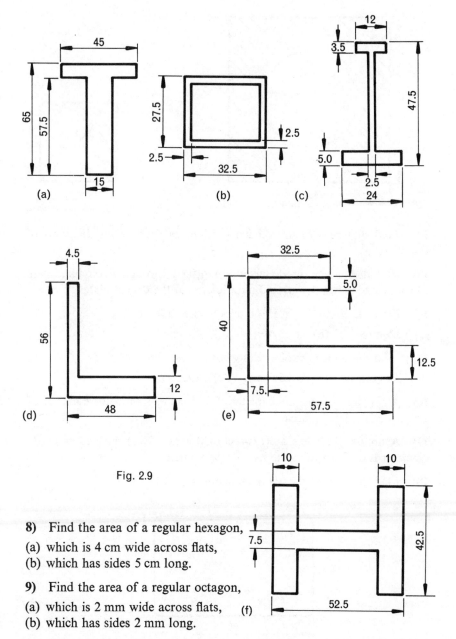

Fig. 2.9

8) Find the area of a regular hexagon,

(a) which is 4 cm wide across flats,
(b) which has sides 5 cm long.

9) Find the area of a regular octagon,

(a) which is 2 mm wide across flats,
(b) which has sides 2 mm long.

10) The parallel sides of a trapezium are 12 cm and 16 cm long. If its area is 220 cm² what is its altitude?

11) If the area of cross-section of a circular shaft is 7 cm² find its diameter.

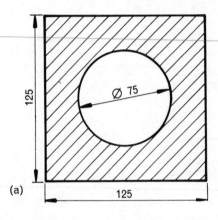

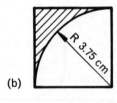

(a)

(b)

Fig. 2.10

12) Find the areas of the shaded portions of each of the diagrams of Fig. 2.10.

13) 10 holes, 5 mm in diameter are drilled through a circular plate which has a 90 mm diameter. Find the area of the drilled plate.

14) Find the circumference of circles whose radii are:

(a) 3.5 mm   (b) 13.8 m   (c) 4.2 cm

15) Find the diameters of circles whose circumferences are:

(a) 34.4 mm   (b) 18.54 cm   (c) 195.2 m

16) A ring has an outside diameter of 3.85 cm and an inside diameter of 2.63 cm. Calculate its area.

17) A hollow shaft has a cross-sectional area of 8.68 cm². If its inside diameter is 0.75 cm calculate its outside diameter.

18) Find the area of the blank shown in Fig. 2.11.

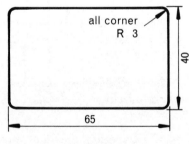

Fig. 2.11

19) How many revolutions will a wheel make in travelling 2 km if its diameter is 700 mm?

20) If r is the radius and θ is the angle subtended by an arc, find the length of arc when:
(a) $r = 2$ cm, $\theta = 30°$,
(b) $r = 3.4$ cm, $\theta = 38° 40'$.

21) If l is the length of an arc, r is the radius and θ the angle subtended by the arc, find θ when:
(a) $l = 9.4$ cm, $r = 4.5$ cm,
(b) $l = 14$ mm, $r = 79$ mm.

22) If an arc 7 cm long subtends an angle of 45° at the centre, what is the radius of the circle?

23) Find the areas of the following sectors of circles:
(a) radius 3 m, angle of sector 60°
(b) radius 2.7 cm, angle of sector 79° 45'
(c) radius 7.8 cm, angle of sector 143° 42'

24) Calculate the area of the part shaded in Fig. 2.12.

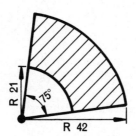

Fig. 2.12

25) A chord 2.6 cm is drawn in a circle of 3.5 cm diameter. What are the lengths of arcs into which the circumference is divided?

26) The radius of a circle is 60 mm. A chord is drawn 40 mm from the centre. Find the area of the minor segment.

27) In a circle of radius 3 cm a chord is drawn which subtends an angle of 80° at the centre. What is the area of the minor segment?

28) A flat is machined on a circular bar of 15 mm diameter, the maximum depth of cut being 2 mm. Find the area of the cross-section of the finished bar.

VOLUMES AND SURFACE AREAS

Any solid having a uniform cross-section and parallel end faces

Volume = Cross-sectional area × length of solid.

Surface Area = Longitudinal surface+ends i.e. (perimeter of cross-section × length of solid) + (total area of ends).

A Prism is the name often given this type of solid if the cross-section is triangular or polygonal.

EXAMPLE 15

An iron bar has the shape of a prism, the cross-section being a parallelogram as shown in Fig. 2.13. If its length is 30 cm find:

a) its volume,

b) its total surface area.

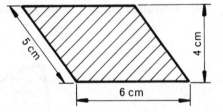

Fig. 2.13

a) Volume = cross-sectional area × length

$$= (6 \times 4) \times 30$$

$$= 720 \text{ cm}^3$$

b) Total surface area

$$= \text{perimeter of cross-section} \times \text{length} + \text{area of ends}$$

$$= (6+6+5+5) \times 30 + 2(6 \times 4)$$

$$= 660 + 48$$

$$= 708 \text{ cm}^2$$

EXAMPLE 16

A steel section has the cross-section shown in Fig. 2.14. If it is 9 m long calculate its volume and total surface area.

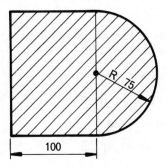

Fig. 2.14

To find the volume:

$$\text{Area of cross-section} = \tfrac{1}{2} \times \pi \times 75^2 + 100 \times 150$$

$$= 23\ 836 \text{ mm}^2$$

$$= \frac{23\ 836}{(1000)^2} = 0.023\ 836 \text{ m}^2$$

∴ $$\text{Volume of solid} = 0.023\ 836 \times 9$$

$$= 0.214\ 5 \text{ m}^3$$

To find the surface area:

$$\text{Perimeter of cross-section} = \pi \times 75 + 2 \times 100 + 150$$

$$= 585.5 \text{ mm}$$

$$= \frac{585.5}{1000} = 0.585\ 5 \text{ m}$$

$$\text{lateral surface area} = 0.585\ 5 \times 9 = 5.270 \text{ m}^2$$

$$\text{surface area of ends} = 2 \times 0.024 = 0.048 \text{ m}^2$$

∴ $$\text{Total surface area} = 5.270 + 0.048$$

$$= 5.318 \text{ m}^2$$

Cylinder

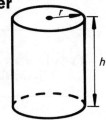

$$\text{Volume} = \pi r^2 h$$

$$\text{Surface area} = 2\pi r h + 2\pi r^2 = 2\pi r(h+r)$$

EXAMPLE 17

A cylindrical can holds 18 litres of petrol. Find the depth of the petrol if the can has a diameter of 60 cm.

Now   18 litres = 18×1000 = 18 000 cm³

and if the depth of the petrol is h cm

then   volume of petrol = $\pi(\text{radius})^2 \times h$

∴ $18\,000 = \pi \times 30^2 \times h$

∴ $h = \dfrac{18\,000}{\pi \times 900}$

                    $= 6.37$ cm

EXAMPLE 18

A metal bar of length 200 mm and diameter 75 mm is melted down and cast into washers 2.5 mm thick with an internal diameter of 12.5 mm and external diameter 25 mm. Calculate the number of washers obtained assuming no loss of metal.

Volume of original bar of metal $= \pi \times 37.5^2 \times 200$

$= 883\,500$ mm³

Volume of one washer $= \pi \times (12.5^2 - 6.25^2) \times 2.5$

$= \pi \times 117.2 \times 2.5$

$= 920.4$ mm³

Number of washers obtained $= \dfrac{883\,500}{920.4}$

$= 960$

Cone

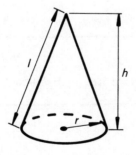

Volume $= \tfrac{1}{3}\pi r^2 h$
(h is the vertical height)

Curved surface area $= \pi r l$
(l is the slant length)

EXAMPLE 19

A right circular cone has a base diameter equal to its height. Find these dimensions if the volume of the cone is 150 cm³. Calculate also the total surface area of the cone.

We are given that $h = 2r$

but $\text{Volume} = \frac{1}{3}\pi r^2 h$

$\therefore$ $150 = \frac{1}{3}\pi \times r^2 \times 2r$

$\therefore$ $r^3 = \frac{3 \times 150}{2\pi} = 71.62$

$\therefore$ $r = 4.15 \text{ cm}$

Hence the diameter and height are 8.30 cm.

The slant length l may be found using Pythagoras' theorem on the triangle as shown in Fig. 2.15.

$\therefore$ $l^2 = r^2 + h^2$

                       $= 4.15^2 + 8.30^2$

$\therefore$ $l = 9.28 \text{ cm}$

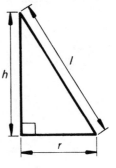

Fig. 2.15

But   total surface area = curved surface area + base area

                       $= \pi r l + \pi r^2$

                       $= \pi \times 4.15 \times 9.28 + \pi \times 4.15^2$

                       $= 121 + 54$

                       $= 175 \text{ cm}^2$

Frustum of a Cone

A *frustum* is the portion of a cone or pyramid between the base and a horizontal slice which removes the pointed portion.

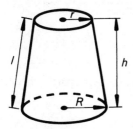

$\text{Volume} = \frac{1}{3}\pi h(R^2 + Rr + r^2)$
(h is the vertical height)

Curved surface area $= \pi l(R+r)$

Total surface area $= \pi l(R+r)+\pi R^2+\pi r^2$

(l is the slant height)

EXAMPLE 20

A taper piece has diameters at its ends of 8 cm and 6 cm respectively and it is 9 cm long. Find its volume.

The figure is a frustum of a cone in which we have $h = 9$ cm, $R = 4$ cm and $r = 3$ cm.

Volume of frustum of cone

$$= \tfrac{1}{3}\pi h\{R^2+rR+r^2\}$$

$$= \frac{\pi \times 9}{3}\cdot\{4^2+3\times4+3^2\}$$

$$= 3\pi\{16+12+9\}$$

$$= 3\pi\times37$$

$$= 348 \text{ cm}^3$$

EXAMPLE 21

The bowl shown in Fig. 2.16 is made from sheet steel and has an open top. Calculate the total cost of painting the vessel (inside and outside) at a cost of 1p per 100 cm².

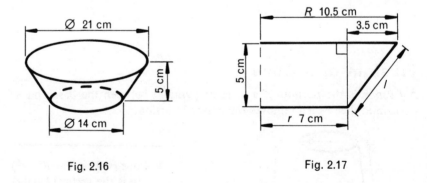

Fig. 2.16   Fig. 2.17

Fig. 2.17 shows a half section of the bowl. Using Pythagoras' theorem on the right angled triangle

$$l^2 = 5^2 + 3.5^2$$

∴ $l = 6.10$ cm

Now the required total surface area, i.e. inside and outside

$$= 2\{\text{curved surface area}\} + 2(\text{base area})$$

$$= 2\{\pi l(R+r)\} + 2(\pi r^2)$$

$$= 2\{\pi(6.10)(10.5+7)\} + 2(\pi 7^2)$$

$$= 670 + 308$$

$$= 978 \text{ cm}^2$$

@ 1p per 100 cm² total cost $= \dfrac{978}{100}$ p

$$= 9.78 \text{ p}$$

Sphere

Volume $= \frac{4}{3}\pi r^3$

Surface area $= 4\pi r^2$

EXAMPLE 22

The flask shown in Fig. 2.18 is completely filled with oil whose density is 0.8 g/cm³. Find the mass of oil in the flask.

Volume of the sphere

$$= \frac{4}{3}\pi \times 25^3 = 65\,450 \text{ cm}^3$$

Volume of the cylinder

$$= \pi \times 1.5^2 \times 20 = 141 \text{ cm}^3$$

Volume of the flask

$$= 65\,450 + 141 = 65\,591 \text{ cm}^3$$

∴ mass of the oil $= $ volume $\times$ density

$$= 65\,591 \times 0.8$$

$$= 52\,470 \text{ g}$$

$$= 52.47 \text{ kg}$$

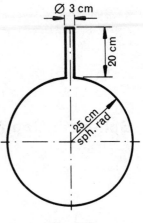

Fig. 2.18

EXAMPLE 23

A spherical fuel container has an outer surface area of 10.5 m². Neglecting the thickness of the spherical wall find:

a) the diameter of the sphere;

b) the capacity of the sphere in litres.

a)   Now the surface area of a sphere $= 4\pi(\text{radius})^2$

$\therefore$ $10.5 = 4\pi(\text{radius})^2$

$\therefore$ $\text{radius} = \sqrt{\dfrac{10.5}{4 \times \pi}}$

$$= 0.914 \text{ m}$$

b) Since the capacity of the sphere is equal to its volume

then $\text{capacity} = \dfrac{4}{3}\pi(\text{radius})^3$

$$= \dfrac{4}{3}\pi(0.914)^3$$

$$= 3.198 \text{ m}^3$$

$$= 3.198 \times 1000 = 3198 \text{ litres}$$

EXAMPLE 24

A right circular cone has the same base radius r as the radius of a sphere and its volume is twice that of the sphere. Calculate the vertical height of the cone in terms of r.

$$\text{Volume of sphere} = \tfrac{4}{3}\pi r^3$$

If h = height of cone, then:

$$\text{Volume of cone} = \tfrac{1}{3}\pi r^2 h$$

$\therefore$ $\tfrac{1}{3}\pi r^2 h = 2 \times \tfrac{4}{3}\pi r^3$

$\therefore$ $\pi r^2 h = 8\pi r^3$

$\therefore$ $h = 8r$

Spherical Cap (i.e. segment of a sphere)

$$\text{Volume} = \frac{\pi h^2}{3}(3R - h)$$

$$\text{or} \quad \frac{\pi h}{6}(3r^2 + h^2)$$

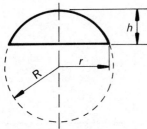

$$\text{Surface area} = \text{Curved surface area} + \text{flat base area}$$
$$= 2\pi rh + \pi h(2r - h)$$
$$= \pi h\,(4r - h)$$

EXAMPLE 25

A casting is the shape of an inverted frustum of a cone, hollowed out in a spherical shape as shown in Fig. 2.19. Calculate the base radius x cm of the spherical segment if the casting volume is 15 cm³.

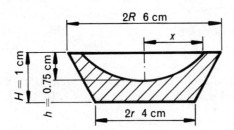

Fig. 2.19

Now,

$$\text{vol. of casting} = \text{vol. of frustum} - \text{vol. of segment}$$
$$= \frac{\pi H}{3}(R^2 + rR + r^2) - \frac{\pi h}{6}\left(3x^2 + h^2\right)$$
$$= \frac{\pi 1}{3}(3^2 + 2\times 3 + 2^2) - \frac{\pi 0.75}{6}\left(3x^2 + (0.75)^2\right)$$

$$\therefore \qquad 15 = \frac{19\pi}{3} - \frac{\pi}{8}\left(3x^2 + 0.56\right)$$

$$\therefore \qquad \frac{\pi}{8}\left(3x^2 + 0.56\right) = \frac{19\pi}{3} - 15 = 4.9$$

$$\therefore \quad 3x^2 = \frac{4.9 \times 8}{\pi} - 0.56$$

$$= 12.5 - 0.56$$

$$= 11.94$$

$$\therefore \qquad x^2 = 3.98$$

$$\therefore \qquad x = 2 \text{ cm} \quad \text{(almost)}.$$

Pyramid

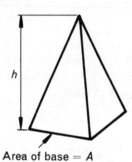

Area of base = A

Volume $= \frac{1}{3}Ah$

Surface area = Sum of the areas of the triangles forming the sides plus the area of the base.

EXAMPLE 26

Find the volume and total surface area of a symmetrical pyramid whose base is a rectangle 7 m × 4 m and whose height is 10 m.

$$\text{Area of base} = 7 \times 4 = 28 \text{ m}^2$$

$$\text{height} = 10 \text{ m}$$

$$\therefore \qquad \text{volume} = \frac{1}{3}Ah = \frac{1}{3} \times 28 \times 10$$

$$= 93.3 \text{ m}^3$$

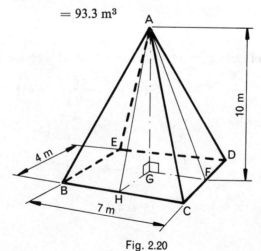

Fig. 2.20

From Fig. 2.20 the surface area consists of two sets of equal triangles (that is $\triangle ABC$ and $\triangle ADE$, and also $\triangle ABE$ and $\triangle ACD$) together with the base BCDE. To find the area of $\triangle ABC$ we must find the slant height AH. From the apex A drop a perpendicular AG onto the base and draw GH perpendicular to BC. H is then the mid-point of BC.

In $\triangle AHG$, $\angle AGH = 90°$ and, by Pythagoras' Theorem,

$$AH^2 = AG^2 + HG^2 = 10^2 + 2^2$$

$$= 100 + 4 = 104$$

$\therefore$ $\qquad AH = \sqrt{104} = 10.20 \text{ m}$

$\therefore$ $\qquad$ area of $\triangle ABC = \frac{1}{2} \times \text{base} \times \text{height}$

$$= \frac{1}{2} \times 7 \times 10.20 = 35.7 \text{ m}^2$$

Similarly, to find the area of $\triangle ACD$ we must find the slant height AF. Draw GF, F being the mid-point of CD. Then in $\triangle AGF$, $\angle AGF = 90°$ and by Pythagoras' Theorem,

$$AF^2 = AG^2 + GF^2 = 10^2 + 3.5^2$$

$$= 100 + 12.25 = 112.25$$

$\therefore$ $\qquad AF = \sqrt{112.25} = 10.59 \text{ m}$

$\therefore$ $\qquad$ area of $\triangle ACD = \frac{1}{2} \times \text{base} \times \text{height}$

$$= \frac{1}{2} \times 4 \times 10.59 = 21.18 \text{ m}^2$$

$\therefore$ $\qquad$ total surface area $= (2 \times 35.7) + (2 \times 21.18) + (7 \times 4)$

$$= 71.4 + 42.36 + 28$$

$$= 141.8 \text{ m}^2$$

Frustum of a Pyramid

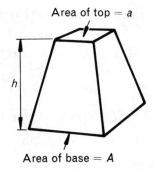

Area of top = a

h

Area of base = A

Volume $= \frac{1}{3}h(A + \sqrt{Aa} + a)$

Surface area = Sum of the areas of the trapeziums forming the sides plus the areas of the top and base of the frustum.

EXAMPLE 27

A casting has a length of 2 m and its cross-section is a regular hexagon. The casting tapers uniformly along its length, the hexagon having a side of 200 mm at one end and 100 mm at the other. Calculate the mass of the casting if the material has density of 7.8 g/cm³.

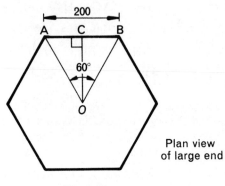

Fig. 2.21

In Fig. 2.21 the area of the hexagon of 200 mm side

$$= 6 \times \text{area } \triangle ABO$$

In $\triangle AOC$, $\angle AOC = 30°$, $AC = 100$ mm

$$\therefore \qquad OC = \frac{AC}{\tan 30°}$$

$$= \frac{100}{\tan 30°} = 173.2 \text{ mm}$$

$$\therefore \qquad \text{Area of } \triangle ABO = \tfrac{1}{2} \times 200 \times 173.2$$

$$\therefore \qquad \text{area of hexagon} = 6 \times \tfrac{1}{2} \times 200 \times 173.2 = 103\,920 \text{ mm}^2$$

The area of the hexagon of 100 mm side can be found in the same way. It is 25 980 mm².

The casting is a frustum of a pyramid with $A = 103\,920$, $a = 25\,980$ and $h = 2000$

$$\therefore \qquad \text{volume} = \tfrac{1}{3}h(A + \sqrt{Aa} + a)$$

$$= \tfrac{1}{3} \times 2000 \times (103\,920 + \sqrt{103\,920 \times 25\,980} + 25\,980)$$

$$= \tfrac{1}{3} \times 2000 \times (103\,920 + 51\,960 + 25\,980)$$

$$= 12.124 \times 10^7 \text{ mm}^3$$

but   mass of casting $=$ volume $\times$ density

$$= 12.124 \times 10^7 \times \frac{7.8}{1000}$$

$$= 945\,672 \text{ g}$$

$$= 945.7 \text{ kg}$$

Exercise 11

1) A steel ingot whose volume is 2 m³ is rolled into plate 15 mm thick and 1.75 m wide. Calculate the length of the plate in metres.

2) A block of lead 1.5 m $\times$ 1 m $\times$ 0.75 m is hammered out to make a square sheet 10 mm thick. What are the dimensions of the square?

3) Calculate the volume of a metal tube whose bore is 50 mm and whose thickness is 8 mm if it is 6 m long.

4) The volume of a small cylinder is 180 cm³. If the diameter of the cross-section is 25 mm find its height.

5) A steel ingot is in the shape of a cylinder 1.5 m diameter and 3.5 m long. How many metres of square bar of 50 mm side can be rolled from it.

6) A metal cone has a diameter of 70 mm and a height of 100 mm. What is its volume?

7) Calculate the diameter of a tin can whose height is the same as its diameter and whose volume is 220 cm³.

8) A piece of alloy with dimensions 20 mm $\times$ 60 mm $\times$ 1800 mm is melted down and recast into a cylinder whose diameter is 140 mm. Find the length of the cylinder.

9) It is required to replace two pipes with internal diameters 28 mm and 70 mm by a single pipe which will provide an equal area of flow. Calculate the internal diameter of pipe required.

10) An ingot whose volume is 2 m³ is to be made into ball bearings whose diameters are 12 mm. Assuming 20% of the metal in the ingot is wasted how many ball bearings will be produced?

11) Find the volume of a hexagonal bar which is 4 cm across flats and 3 m long.

12) A hemi-spherical bowl has an external diameter of 12 cm and an internal diameter of 11 cm. It is made of copper which has a density of 8.9 g/cm³. What is the mass of the bowl?

13) Find the volumes of the bars shown in cross-section in Fig. 2.22. Each bar is 10 m long. Give the answers in m³.

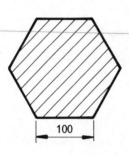

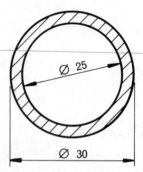

Fig. 2.22

14) Find the volumes and surface area of the following solids:

(a) a cone with a diameter of 6 cm and a height of 5 cm,

(b) a sphere with a diameter of 2.5 cm.

15) A hemispherical bowl has an external diameter of 12 cm and an internal diameter of 11 cm. It is made of copper which has a mass of 8.9 gram per cubic centimetre. Find the mass of the bowl.

16) The washer shown in Fig. 2.23 has a square of side l cut out of it. If its thickness is t, find an expression for the volume, V, of the washer. Hence find the volume of a washer when $D = 6$ cm, $t = 0.2$ cm and $l = 4$ cm.

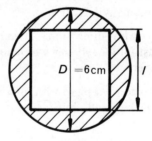

Fig. 2.23

17) A cylindrical tank 4 m high and open at the top has a radius of 2 m. Find its capacity in litres.

18) A hollow shaft has an outside diameter of 80 mm. It has the same mass as a solid shaft of the same length and material which is 40 mm in diameter. What is the thickness of the metal in the hollow shaft?

19) A pyramid has a square base of side 2 cm and height 4 cm. Find its volume and total surface area.

20) A pyramid has a base which is an octagon of 30 mm side. It has a height of 80 mm. Find the volume and surface area of the pyramid.

21) The end section of a motor shaft is in the form of the major arc of a circle of radius 5 cm cut off by a chord of length 4 cm. Calculate the area of the section.

22) Find the diameter of the largest sphere which can be placed inside a hollow cone whose height and internal base diameter are both 50 cm. No part of the sphere is to protrude from the cone. Calculate also the internal volume and surface area of the cone.

23) A metal bucket with an open top is in the shape of the frustum of a cone. If its top diameter is 30 cm, its bottom diameter is 20 cm and its height is 40 cm find its capacity in litres. Find also the area of metal used in its construction neglecting any overlaps, etc.

24) A frustum of a cone has a volume of 12 400 cm^3. If its large diameter is 32 cm and its small diameter 24 cm find its height.

25) A waste bin is made from sheet metal in the form of a frustum of a pyramid. The open top is a square of side 25 cm, the bottom is a square of side 20 cm and the height is 30 cm. Find:

(a) its capacity,
(b) the total surface area to be painted (inside and outside).

NUMERICAL METHODS FOR CALCULATING IRREGULAR AREAS AND VOLUMES

AREAS

An irregular area is one whose boundary does not follow a definite pattern, e.g. the cross-section of a river.

In these cases practical measurements are made and the results plotted to give a graphical display.

Various numerical methods may then be used to find the area.

USE OF GRAPH PAPER AND 'COUNTING THE SQUARES'

This is probably the simplest method and is often overlooked. It gives results as accurately as those obtained by other more complicated methods.

EXAMPLE 28

A series of soundings taken across the section of a river channel are shown in Fig. 2.24, all dimensions being metres. Find the area of the river cross-section.

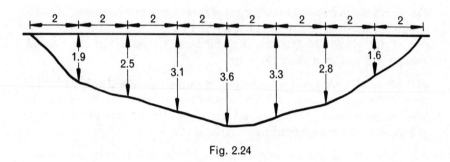

Fig. 2.24

The scales are often chosen to enable the diagram to fit conveniently on a particular size of paper available — the larger the diagram the better.

Fig. 2.25 shows the given results and the plotted points have been joined with a reasonably smooth curve (this is preferable to joining the points with straight lines).

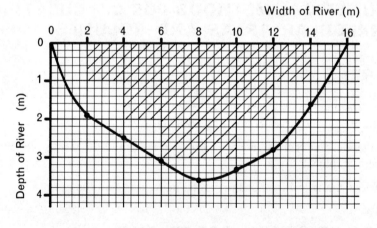

Fig. 2.25

The number of *whole* large squares are then counted — these are shown shaded. The remainder of the area is found by counting the smaller squares, judgement being made for portions of these smaller squares.

Results   Number of large squares = 12

Number of small squares = 173

But on the paper used there are 25 small squares per each large square.

Hence the total number of large squares = $12 + \dfrac{173}{25}$

$$= 18.9$$

The scales of the axes must be taken into account:

Consider one large square as shown in Fig. 2.26.

The area of one large square

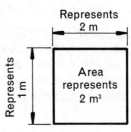

= 2 m (horizontal)

× 1 m (vertical)

= 2 m²

Hence the required area of the river cross-section

= 18.9 × 2 m²

= 37.8 m²

Fig. 2.26

Rough check of result: As in all engineering calculations a rough check should always be made to ensure that the answer obtained is reasonable. This helps to avoid making a big mistake — for instance in this example we may have forgotten to multiply by the scale factor of 2 and obtained an answer one half of the correct value.

The cross-sectional area is approximately equal to that of the rectangle shown in Fig. 2.27,

i.e. 16 m × 2.2 m = 35.2 m²

This confirms that the answer obtained is of a reasonable magnitude.

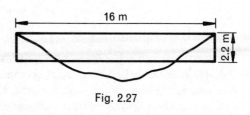

Fig. 2.27

Accuracy of result: The accuracy of the original measurements, the choice of profile when joining the plotted points, and the counting of the incomplete small squares will all affect the final result. This is a difficult example in which to try and obtain a mathematically calculated error. This will be done in examples which follow later in this chapter. From experience, however, we would expect the result obtained to be within an accuracy of ± 5%.

THE PLANIMETER

Areas may be measured by means of an instrument called a *planimeter* (Fig. 2.28). It consists of two bars A and B which are hinged at C. Bar B can rotate about a needle point pushed into the paper at D. D is loaded with a weight W to prevent the needle jumping out. Arm A rests on a wheel E, which may roll on the paper as A is moved, and also on an adjustable foot F next to the tracing needle at T.

The tracing point T is taken round the perimeter of the area to be measured and the area read off on the scale attached to wheel E.

The scale has 100 divisions together with an adjacent vernier which enables the scale to be read to one tenth of a division. A small indicator wheel G registers the number of complete revolutions of wheel E (just as the hour hand of a clock records the number of complete revolutions which the minute hand makes).

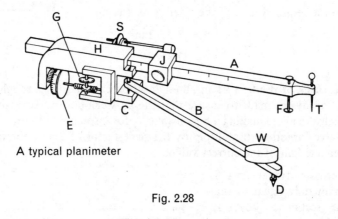

A typical planimeter

Fig. 2.28

The wheel assembly is carried by a portion H whose position is fixed by sliding along bar A before measuring commences. This position is chosen by deciding the units of the required area, e.g. square centimetres, etc. and setting according to the instructions inside the lid of the planimeter case. The instrument should be used on a sheet of drawing paper perferably in a horizontal plane which is sufficiently large to enable the whole of the movement of the wheel E to be made without coming off the paper. The surface of the paper should not be highly polished, otherwise skidding of the wheel may occur (other than that intended) whilst the needle point is being taken round the area perimeter, and hence an incorrect measurement may result.

The tracing point should always be taken clockwise round the boundary —

any point may be chosen for the start but the tracing point must always be brought back to that point to complete a measurement. It is best to arrange the initial position so that the arms A and B are approximately at right angles.

This all sounds very complicated but a few minutes using the instrument will show you that it is really quite easy to use, and although you may feel that your hand is not causing the needle to follow the boundary line exactly the errors on each side will usually cancel each other out and a surprisingly accurate measurement can be made after some practice.

An example on the use of the planimeter

Draw a square of side 5 cm. Next set the sliding portion H to the position for measuring square centimetres (see inside the lid of the case). Clamp the small carrier J to the arm A with the pointer on H in approximately the correct position and then set it accurately using the fine adjusting screw S.

Adjust the screw foot F so that it takes the weight of the arm of the tracing needle and prevents it catching and sticking in the surface of the paper. Locate the needle point at the spot on the perimeter which you have chosen for the start and take the scale reading. Trace clockwise round the boundary of the square and again take the scale reading when the whole of the boundary has been covered and the needle is back again at the starting point. The difference between these two readings gives the required area, and should of course give 25 cm². Check by a second circuit.

MID-ORDINATE RULE

Suppose we wish to find the area shown in Fig. 2.29. Let us divide the area into a number of vertical strips, each of equal width b.

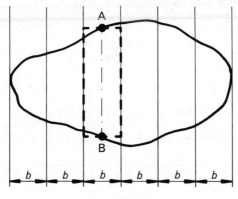

Fig. 2.29

Consider the 3rd strip, whose centre-line is shown cutting the curved boundaries of the area at A and B respectively. Through A and B horizontal lines are drawn and these help to make the dotted rectangle shown. The rectangle has approximately the same area as that of the original 3rd strip, and this area will be $b \times AB$.

AB is called the mid-ordinate of the 3rd strip, as it is mid way between the vertical sides of the strip.

To find the *whole area*, the areas of the other strips are found in a similar manner and then all are added together for the final result.

$$\therefore \qquad \text{Area} = \text{width of strip} \times \text{sum of the mid-ordinates.}$$

A useful practical tip to avoid measuring each separate mid-ordinate is to use a strip of paper and mark off along its edge successive mid-ordinate lengths, as shown in Fig. 2.30. The total area will then be found by measuring the whole length marked out (in the case shown this is HR) and multiplying by the strip width b.

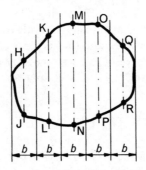

Fig. 2.30

EXAMPLE 29

Find the area under the curve $y = x^2 + 2$ between $x = 1$ and $x = 4$.

The curve is sketched in Fig. 2.31. Taking 6 strips we may calculate the mid-ordinates.

x	1.25	1.75	2.25	2.75	3.25	3.75
y	3.56	5.06	7.06	9.56	12.56	16.06

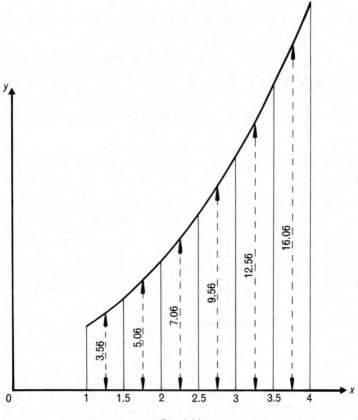

Fig. 2.31

Since the width of the strips $= \frac{1}{2}$, the mid-ordinate rule gives:

$$\text{Area} = \tfrac{1}{2} \times (3.56 + 5.06 + 7.06 + 9.56 + 12.56 + 16.06)$$
$$= \tfrac{1}{2} \times 53.86 = 26.93 \text{ square units}$$

It so happens that in this example it is possible to calculate an exact answer. How this is done need not concern us at this stage, but by comparing the exact answer with that obtained by the mid-ordinate rule we can see the size of the error.

$$\text{Exact answer} = 27 \text{ square units}$$

Approximate answer (using the mid-ordinate rule)

$$= 26.93 \text{ square units}$$

$\therefore$ $$\text{Error} = 0.07 \text{ square units}$$

$\therefore$ $$\text{Percentage error} = \frac{0.07}{27} \times 100 = 0.26\%$$

From the above it is clear that the mid-ordinate rule gives a good approximation to the correct answer.

TRAPEZOIDAL RULE

Consider the area having boundary ABCD shown in Fig. 2.32.

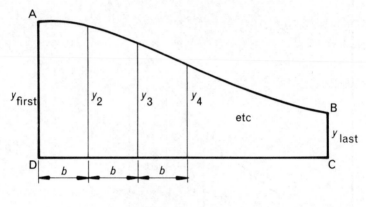

Fig. 2.32

The area is divided into a number of vertical strips of equal width b.

Each vertical strip is assumed to be a trapezium. Hence the third strip, for example, will have an area $= b \times \frac{1}{2}(y_3+y_4)$.

But   Area ABCD $=$ the sum of all the vertical strips

$$= b \times \tfrac{1}{2}(y_{first}+y_2)+b \times \tfrac{1}{2}(y_2+y_3)+b \times \tfrac{1}{2}(y_3+y_4)+ \ldots$$
$$= b[\tfrac{1}{2}y_{first}+\tfrac{1}{2}y_2+\tfrac{1}{2}y_2+\tfrac{1}{2}y_3+ \ldots +\tfrac{1}{2}y_{last}]$$
$$= b[\tfrac{1}{2}(y_{first}+y_{last})+y_2+y_3+y_4 \ldots]$$
$$= \text{width of strips} \times [\tfrac{1}{2}(\text{sum of the first and last ordinates})$$
$$+(\text{the sum of the remaining oridnates})]$$

The accuracy of the trapezoidal rule is similar to that of the mid-ordinate rule. A comparison may be made by solving Example 29 using the trapezoidal rule.

EXAMPLE 30

Find the area under the curve $y = x^2+2$ between $x = 1$ and $x = 4 \cdot$

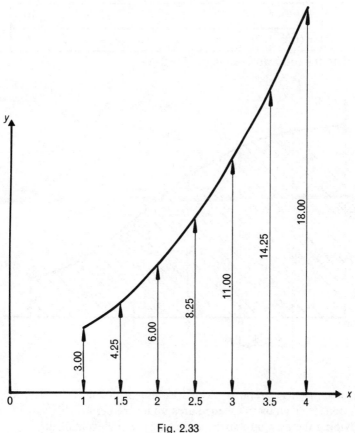

Fig. 2.33

The curve is sketched in Fig. 2.33, the lengths of the ordinates having been calculated.

Since the width of the strips $= \frac{1}{2}$, the trapezoidal rule gives:

$$\text{Area} = \tfrac{1}{2} \times [\tfrac{1}{2}(3+18)+4.25+6+8.25+11+14.25]$$
$$= \tfrac{1}{2} \times [10.5+43.75]$$
$$= 27.13 \text{ square units.}$$

The exact answer is 27 square units and therefore:

$$\text{Percentage error} = \frac{27.13-27}{27} \times 100 = 0.48\%$$

EXAMPLE 31

The table gives the values of a force required to pull a trolley when measured at various distances from a fixed point in the direction of the force:

F (N)	51	49	45	37	26	15	10
s (m)	0	1	2	3	4	5	6

Calculate the total work done by this force.

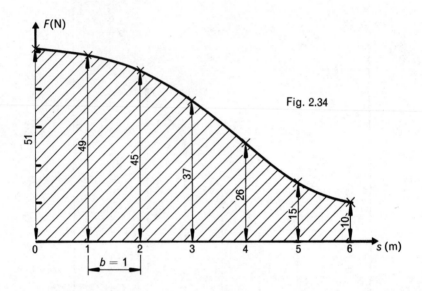

Fig. 2.34

The force-distance graph is plotted as shown in Fig. 2.34. The required work done is given by the shaded area under the curve. This may be found by dividing the area into strips and using the trapezoidal rule.

$\therefore \qquad$ Area $= 1[\frac{1}{2}(51+10)+(49+45+37+26+15)]$

$\qquad\qquad\quad = 203$

$\therefore \qquad$ Work done $= 203$ Nm

$\qquad\qquad\qquad = 203$ J   (since joule $=$ newton $\times$ metre)

SIMPSON'S RULE

The required area is divided into an *even* number of vertical strips of equal width b.

Then Simpson's rule gives:

$$\text{Area} = \frac{b}{3}[(\text{the sum of the first and last ordinates})$$
$$+2(\text{the sum of the remaining odd ordinates})$$
$$+4(\text{the sum of the even ordinates})]$$

This rule usually gives a more accurate result than either the mid-ordinate or trapezoidal rules, but is slightly more complicated to use.

N.B. there must be an *even* number of strips.

EXAMPLE 32

The velocity of a car, starting from rest, is given by the following table:

Time s	0	10	20	30	40	50	60
Velocity m/s	0	16	28	36	40	42	43

Draw a graph showing the relationship between velocity and time, and hence find the distance travelled in 60 s from rest.

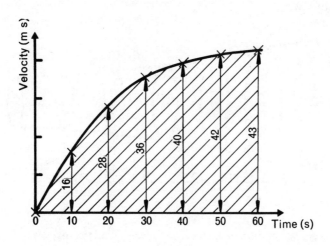

Fig. 2.35

The area under the velocity-time graph gives the distance travelled and it is shown shaded in Fig. 2.35.

Simpson's rule gives:

$$\text{Area} = \tfrac{10}{3} \, [(0+43)+4(16+36+42)+2(28+40)]$$

$$= 1850$$

The units of the answer are:

$$\frac{m}{s} \, (\text{vert.}) \times s \, (\text{horiz.}) = m$$

∴   distance travelled = 1850 m.

VOLUMES OF SOLIDS

All the methods explained in this chapter for finding irregular areas may be applied to finding volumes. The following example shows a typical problem solved by the use of Simpson's rule and the trapezoidal rule.

EXAMPLE 33

The diameters in metres of a felled tree trunk at one metre intervals along its length are as follows:

$$1.00, \ 0.90, \ 0.81, \ 0.74, \ 0.68, \ 0.64 \ \text{and} \ 0.61$$

Assuming that the cross-sections of the trunk are circular estimate the volume of timber.

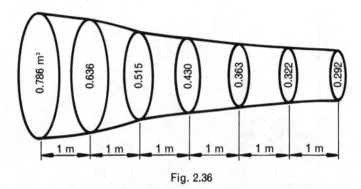

Fig. 2.36

The areas corresponding to each diameter are calculated using the formula:

Area $= \dfrac{\pi}{4}$ (diameter)^2, and are as shown in Fig. 2.36.

The graph of area against length could be plotted and the cross-sectional areas would be the lengths of ordinates. The area under this curve would then represent volume.

Using Simpson's rule we have

Volume $= \dfrac{b}{3}$ [(the sum of the first and last ordinates)
    $+2$(the sum of the remaining odd ordinates)
    $+4$(the sum of the even ordinates)]

$= \frac{1}{3}[(0.786+0.292)+2(0.515+0.363)+4(0.636+0.430+0.322)]$

$= 2.79 \ \text{m}^3$ The units being the result of multiplying m² (area) by m (length.)

An alternative layout using a table to show the calculations is often used:

Ordinate number	Ordinate	Simpson's multiplier	Product
1	0.786	1	0.786
2	0.636	4	2.544
3	0.515	2	1.030
4	0.430	4	1.720
5	0.363	2	0.726
6	0.322	4	1.288
7	0.292	1	0.292
		$\therefore$ total product $=$	8.386

Hence volume $= \frac{1}{3}(8.386) = 2.795$ m^3

Using the trapezoidal rule we have

Volume $= b$ [$\frac{1}{2}$(the sum of the first and last ordinates)
        $+$(the sum of the remaining ordinates)]

$= 1[\frac{1}{2}(0.786+0.292)$
  $+(0.636+0.515+0.430+0.363+0.322)]$

$= 2.805$ m^3

These results are reasonably close and we could safely assume that the volume of timber is 2.8 cubic metres.

Exercise 12

It is suggested that the following examples are solved using at least two of the methods covered in the preceding text, i.e. 'counting the squares', using a planimeter, using the trapezoidal rule, using the mid-ordinate rule, or by using Simpson's rule.

1) The table below gives corresponding values of x and y. Plot the graph and by using the mid-ordinate rule find the area under the graph.

x	1.5	1.7	1.9	2.1	2.3	2.5	2.7	2.9	3.1
y	800	730	622	528	438	366	306	262	214

2)   The table below gives corresponding values of two quantities A and x. Draw the graph and hence find the area under it. (Plot x horizontally).

A	53.2	35	22.2	21.8	24.2	23.6	18.7	0
x	0	1	2	3	4	5	6	7

3)   Plot the curve given by the following values of x and y and hence find the area included by the curve and the axes of x and y.

x	1	2	3	4	5
y	1	0.25	0.11	0.063	0.040

4)   Plot the curve of $y = 2x^2 + 7$ between $x = 2$ and $x = 5$ and find the area under this curve.

5)   Plot the graph of $y = 2x^3 - 5$ between $x = 0$ and $x = 3$ and find the area under the curve.

6)   A series of soundings taken across a section of a river channel are given in Fig. 2.37. Find an approximate value for the cross-sectional area of the river at this section.

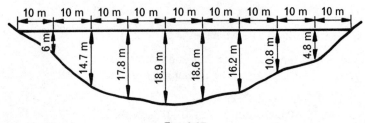

Fig. 2.37

7)   The cross-sectional areas of a tree trunk are given in the table below. Find its volume.

Distance from one end (m)	0	1	2	3	4	5	6
Area (m²)		5.1	4.1	3.4	2.7	2.2	1.8	1.3

8)   The width of a river at a certain section is 60 m. Soundings of the depth of the river taken at this section are recorded as follows:

Distance from L.H. bank (m)	0	5	10	15	20	25	30	35	40	45	50	55	60
Sounding depth (m)	4	8	9	19	30	35	30	24	20	16	10	8	0

Plot the above information and drawing a fair curve through the points, calculate the cross-sectional area of the river at this section in m². If the volume of water flowing past this point per second is 10 000m³, calculate the speed, in m/s, at which the river is flowing.

Hint: The volume of flow per second = (cross-sectional area) × (velocity of flow).

9)   Observation by surveyors show that the cross-sectional areas at 100 m intervals of a cutting are as shown in Fig. 2.38. Find the volume of soil required to fill the cutting.

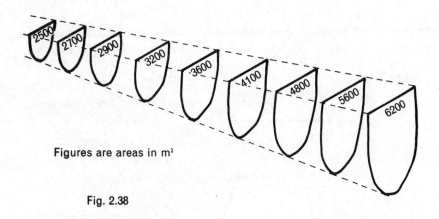

Figures are areas in m²

Fig. 2.38

SUMMARY

Areas and perimeters

The following table gives the areas and perimeters of some simple geometrical shapes.

Figure	Diagram	Formulae
Rectangle		Area $= b \times h$ Perimeter $= 2h + 2b$

Figure	Diagram	Formulae

Parallelogram h Area $= b \times h$

Triangle h Area $= \frac{1}{2} \times b \times h$

Trapezium h Area $= \frac{1}{2} \times h \times (a+b)$

Circle r Area $= \pi r^2$

Circumference $= 2\pi r$

Sector of a circle $\theta°$ Area $= \pi r^2 \times \dfrac{\theta}{360}$

Length of arc $= 2\pi r \times \dfrac{\theta}{360}$

Volumes and surface areas

The following table gives volumes and surface areas of some simple solids.

Figure	Volume	Surface Area
Any solid having a uniform cross-section and parallel end faces.	Cross-sectional area × length of solid.	Curved surface+ends, i.e. (perimeter of cross-section × length of solid)+(total area of ends).

Cylinder

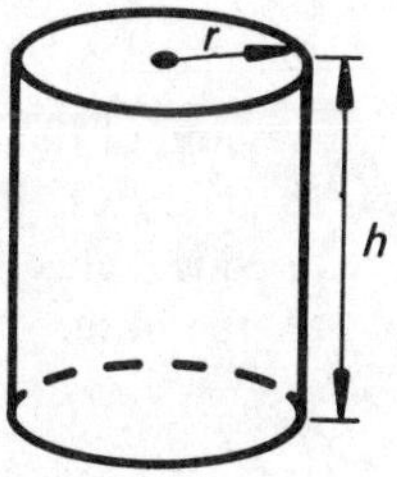

$\pi r^2 h$

$2\pi rh+2\pi r^2 =$
$\qquad\qquad 2\pi r(h+r)$

Cone

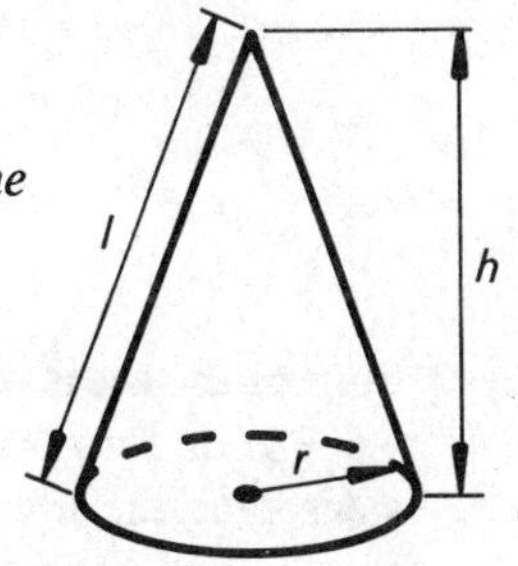

$\frac{1}{3}\pi r^2 h$

(*h* is the vertical height)

πrl

(*l* is the slant height)

Frustum of a cone

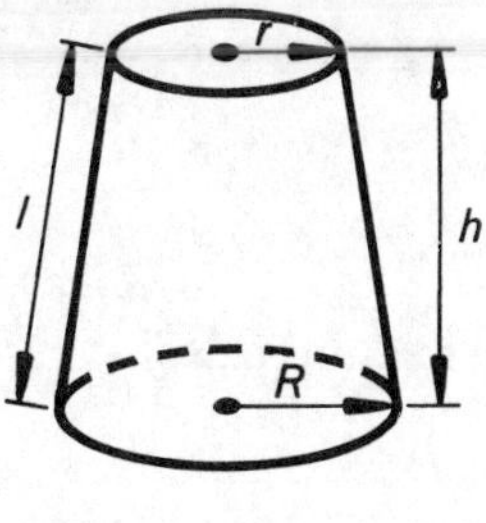

$\frac{1}{3}\pi h(R^2+Rr+r^2)$

(*h* is the vertical height)

Curved surface area =
$\qquad\qquad \pi l(R+r)$

Total surface area =
$\qquad \pi l(R+r)+\pi R^2+\pi r^2$

(*l* is the slant height)

Sphere

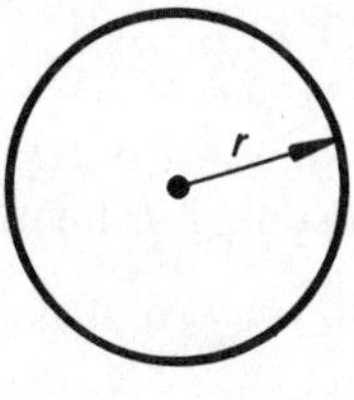

$\frac{4}{3}\pi r^3$

$4\pi r^3$

Figure	Volume	Surface Area
Spherical cap 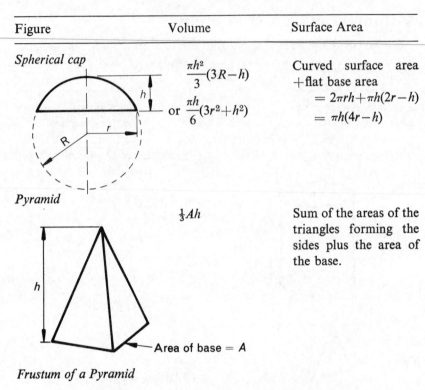	$\dfrac{\pi h^2}{3}(3R-h)$ or $\dfrac{\pi h}{6}(3r^2+h^2)$	Curved surface area +flat base area $= 2\pi rh+\pi h(2r-h)$ $= \pi h(4r-h)$
Pyramid	$\tfrac{1}{3}Ah$	Sum of the areas of the triangles forming the sides plus the area of the base.
Frustum of a Pyramid 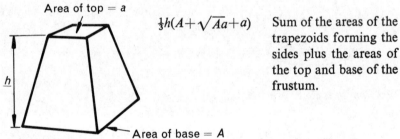	$\tfrac{1}{3}h(A+\sqrt{Aa}+a)$	Sum of the areas of the trapezoids forming the sides plus the areas of the top and base of the frustum.

The mid-ordinate rule

The mid-ordinate rule gives:

Area = width of strip × (the sum of the mid-ordinates)

The trapezoidal rule

The trapezoidal rule gives:

Area = width of strip × [½(sum of the first and last ordinates) +(sum of the remaining ordinates)]

Simpsons rule

Simpson's Rule gives, using an *even* number of strips only:

$$\text{Area} = \frac{\text{width of strip}}{3} \times [(\text{the sum of the first and last ordinates})$$
$$+2(\text{sum of the remaining odd ordinates})$$
$$+4(\text{sum of the even ordinates})]$$

Self Test 2

1) An area is 5 m². Hence it is:

    a 50 cm²   **b** 500 cm²   **c** 5000 cm²   **d** 50 000 cm²

2) An area is 2000 mm². Hence it is:

    a 2 cm²   **b** 20 cm²   **c** 0.2 cm²   **d** 200 cm²

3) An area is 600 000 mm². Hence it is:

    a 6000 m²   **b** 600 m²   **c** 6 m²   **d** 0.6 m²

4) An area is 0.3 m². Hence it is:

    a 30 mm²   **b** 300 mm²   **c** 30 000 mm²   **d** 300 000 mm²

5) The area of a triangle is:

    a base × height   **b** $\frac{1}{2}$ × base × height
    c $\frac{1}{3}$ × base × height   **d** base² × height

6) A triangle has an altitude of 100 mm and a base of 50 mm. Its area is:

    a 2500 mm²   **b** 5000 mm²   **c** 25 cm²   **d** 50 cm²

7) A parallelogram has a base 10 cm long and a vertical height of 5 cm. Its area is:

    a 25 cm²   **b** 50 cm²   **c** 2500 mm²   **d** 5000 mm²

8) A trapezium has parallel sides whose lengths are 18 cm and 22 cm. The distance between parallel sides is 10 cm. Hence the area of the trapezium is:

    a 400 cm²   **b** 200 cm²   **c** 3960 cm²   **d** 495 cm²

9) The area of a circle is given by the formula:

    a $2\pi r^2$ **b** $2\pi r$ **c** πr^2 **d** πr

10) The circumference of a circle is given by the formula:

    a πr^2 **b** $2\pi r$ **c** πr **d** πd

11) A ring has an outside diameter of 8 cm and an inside diameter of 4 cm. Its area is therefore:

    a $\pi(8^2-4^2)$ **b** $8\pi-4\pi$ **c** $\pi(8+4)(8-4)$ **d** $\pi(4^2-2^2)$

12) A wheel has a diameter of 70 cm. The number of revolutions it will make in travelling 55 km is:

    a 2500   **b** 5000   **c** 50 000   **d** 10 000

13) An arc of a circle is 22 cm and the radius of the circle is 140 mm. The angle subtended by the arc is:

    a 90°   **b** 9°   **c** 180°   **d** 18°

14) A sector of a circle subtends an angle of 120°. If the radius of the circle is 42 cm then the area of the sector is:

    a 88 cm²   **b** 1848 cm²   **c** 3696 cm²   **d** 176 cm²

15) A tank has a volume of 8 m³. Hence the volume of the tank is also:

    a 800 cm³   **b** 8000 cm³   **c** 80 000 cm³   **d** 8 000 000 cm³

16) A solid has a volume of 200 000 mm³. Hence the volume of the solid is also:

    a 2000 cm³   **b** 200 cm³   **c** 20 cm³   **d** 20 000 cm³

17) The capacity of a container is 50 litres. Hence its capacity is also:

    a 50 000 cm³   **b** 5000 cm³   **c** 0.5 m³   **d** 0.05 m³

18) The area of the curved surface of a cylinder of radius r and height h is:

    a $2\pi rh$ **b** $2\pi r^2h$ **c** πrh **d** πr^2h

19) The volume of a cylinder of radius r and height h is:

    a $2\pi rh$ **b** $2\pi r^2h$ **c** πrh **d** πr^2h

20) The total surface area of a closed cylinder whose radius is r and whose height is h is:

    a $\pi rh+2\pi r^2$ **b** $\pi r(h+2r)$ **c** $2\pi rh+2\pi r^2$ **d** $2\pi r(h+r)$

21) A small cylindrical container has a diameter of 280 mm and a height of 50 mm. It will hold:

    a 3.08 ℓ **b** 30.8 ℓ **c** 6.16 ℓ **d** 61.6 ℓ

22) The mass of an object is:

    a $\dfrac{\text{density}}{\text{volume}}$ **b** $\dfrac{\text{volume}}{\text{density}}$ **c** volume × density

23) The density of a material is:

    a $\dfrac{\text{volume}}{\text{mass}}$ **b** $\dfrac{\text{mass}}{\text{volume}}$ **c** mass × volume

24) A block of lead has a volume of 880 cm³. If its mass is 80 g, the density of lead is:

    a 0.09 g/cm³   **b** 11 g/cm³   **c** 90 kg/m³   **d** 11 000 kg/m³

25) Water is flowing through a pipe at a speed of 5 m/s. If the bore of the pipe has an area of 2000 cm², the discharge from the pipe per second is:

    a 10 000 cm³   **b** 1 000 000 cm³ **c** 0.010 m³   **d** 1000 ℓ

26) A tank contains 2000 litres of water. It is emptied by means of a pipe through which the water discharges at 4 m³/min. The time taken to empty the tank is:

    a 30 seconds   **b** 500 min   **c** 5 min   **d** 8 min

27) The volume of a cone of base radius r and height h is:

    a $\pi r^2 h$ **b** $2\pi r h$ **c** $\frac{1}{3}\pi r h^2$ **d** $\frac{1}{3}\pi r^2 h$

28) A cone has height of 90 mm and a diameter of 140 mm. Hence, the volume of the cone is:

    a 462 cm³   **b** 19 800 mm³   **c** 462 000 mm³ **d** 19.8 cm³

29) The volume of a sphere of radius r is:

    a $4\pi r^3$ **b** πr^3 **c** $2\pi r^2$ **d** $\frac{4}{3}\pi r^3$

30) The surface area of a sphere of radius r is:

    a $4\pi r^2$ **b** πr^2 **c** $(2\pi r)^2$ **d** $\frac{4}{3}\pi r^3$

31) A test tube whose overall length is h and whose radius is r has a hemi-spherical end. A formula for its volume is:

    a $\pi r^2\left(\frac{2}{3}r + h\right)$ **b** $\pi r^2\left(h - \frac{1}{3}r\right)$ **c** $2\pi r(r + h)$ **d** $\pi r^2(2 + h - r)$

32) The number of strips required for Simpson's rule is:

    a eight   **b** odd   **c** even   **d** odd or even

33) A planimeter is used for measuring:

    a lengths   **b** perimeters   **c** areas   **d** volumes

STATISTICS

VARIABILITY

Variations in size always occur when articles are maufactured. Because of the variability of every manufacturing method the articles produced by a single process will almost certainly differ from each other.

It is not possible, or desirable, to measure manufactured parts to extreme accuracy and measurements which are made are accurate only to the limits imposed by the measuring device. For instance most micrometer readings are made to an accuracy of 0.01 mm.

FREQUENCY DISTRIBUTIONS

Suppose that an inspector measures the lengths of 60 similar components whose nominal length is 125.00 mm. The results might be as follows:

125.00	124.95	125.00	124.98	124.99	125.05	124.97	124.99
125.08	125.00	124.98	125.01	125.02	125.00	125.06	124.98
125.04	125.07	124.96	125.01	124.99	125.02	124.97	125.02
124.98	125.01	124.92	125.01	124.99	125.05	125.01	124.99
125.01	125.04	125.00	124.99	125.03	124.95	125.01	124.96
125.02	124.99	124.97	125.01	124.93	125.02	124.98	125.03
125.00	125.02	124.96	125.00	125.03	125.00	124.99	125.03
124.94	124.99	124.97	124.98				

These figures do not mean very much as they stand and so we arrange them into a frequency distribution. To do this we collect all the 124.92 mm readings together, all the 124.93 mm readings together, and so on. A tally chart (Table 1) is the best way of doing this. Each time a measurement arises a tally mark is placed opposite the appropriate measurement. The fifth tally mark is usually made in an oblique direction thus tying the tally marks into bundles of five, which makes for easier counting. When the tally marks are complete, the marks are counted and the numerical value recorded in the column headed 'frequency', the frequency being the number of times each measurement occurs. From Table 1 it will be seen that the measurement 124.92 mm occurs once (that is it has a frequency of 1), the measurement 125.01 mm occurs eight times (a frequency of 8), and so on.

TABLE 1

Measurement (mm)	Number of parts with this length	Frequency
124.92	1	1
124.93	1	1
124.94	1	1
124.95	11	2
124.96	111	3
124.97	1111	4
124.98	ⅼ𝚮𝚻𝚻 1	6
124.99	ⅼ𝚮𝚻𝚻 1111	9
125.00	ⅼ𝚮𝚻𝚻 111	8
125.01	ⅼ𝚮𝚻𝚻 111	8
125.02	ⅼ𝚮𝚻𝚻 1	6
125.03	1111	4
125.04	11	2
125.05	11	2
125.06	1	1
125.07	1	1
125.08	1	1

THE HISTOGRAM

The frequency distribution becomes even more understandable if we draw a diagram to represent it. The best type of diagram is the histogram (Fig. 3.1) which consists of a set of rectangles each of the same width, whose heights represent the frequencies. On studying the diagram the pattern of the variation is easily understood, most of the values being grouped near the centre of the diagram with a few values more widely dispersed.

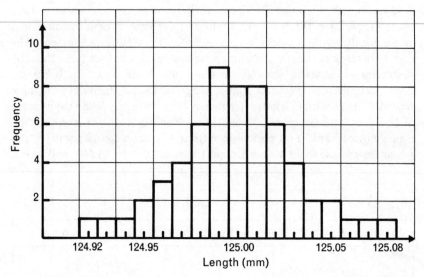

Fig. 3.1

GROUPED DATA

When dealing with a large amount of data it is often useful to group the information into classes or categories. Referring to Table 1, we could re-rearrange the data into a grouped frequency distribution as follows.

TABLE 2

Class	Frequency
124.92–124.94	3
124.95–124.97	9
124.98–125.00	23
125.01–125.03	18
125.04–125.06	5
125.07–125.09	2

The first class consists of measurements between 124.92 and 124.94 mm. Since 3 measurements belong to this class, the class frequency is 3 and similarly for the other classes.

CLASS BOUNDARIES

If we had measured the lengths of the parts in Table 1 to an accuracy of three decimal places then in the first class we would have put all the measurements between 124.915 and 124.945 mm. These measurements are called the lower and upper class boundaries respectively. Similarly for the fourth class the lower class boundary is 125.005 and the upper class boundary is 124.035.

CLASS WIDTH

The width of the class is the difference between the lower and upper class boundaries, that is

Class width = upper class boundary − lower class boundary

For the first class:

Class width = 124.945–124.915 = 0.03 mm

HISTOGRAM FOR A GROUPED DISTRIBUTION

A histogram for a grouped distribution may be drawn by using the lower and upper class boundaries as the extreme values for each rectangle making up the diagram. Strictly, the areas of the rectangles represent the frequencies, but if the class widths are all the same it is usual to make the heights of the rectangles equal to the class frequencies. Fig. 3.2 shows the histogram drawn for the grouped distribution of Table 2.

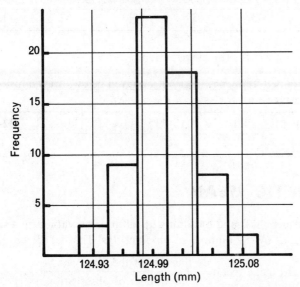

Fig. 3.2

Exercise 13

1)   A patrol inspector visits an automatic lathe once every six minutes. He picks up a component as it drops into the hopper and measures its diameter. After a shift of 4 hours he had collected a sample of 40 components whose diameters are:

24.98	24.96	24.97	24.98	24.99	24.97	25.03
25.00	24.99	25.01	25.03	25.01	25.01	25.00
25.02	25.02	25.00	25.02	25.01	25.04	25.02
25.02	25.01	24.97	24.98	25.01	25.03	24.99
25.03	25.05	24.95	24.98	24.99	25.00	25.01
24.99	25.02	24.97	25.04	25.00		

Draw up a frequency table and hence draw a histogram for these measurements.

2)   The lengths of 80 machined parts are measured with the following results (the measurements are given in hundredths of a millimetre above 62.00 mm).

5, 8, 3, 5, 5, 6, 3, 0, 6, 2, 5, 2, 2, 1, 9, 6, 8, 6, 3, 6, 4, 2, 7, 4, 5, 4, 5, 8, 2, 8, 8, 5, 2, 5, 7, 5, 3, 7, 6, 2, 8, 6, 6, 3, 5, 5, 7, 9, 6, 4, 8, 3, 2, 6, 4, 9, 2, 8, 6, 4, 6, 3, 1, 8, 2, 4, 4, 1, 1, 9, 5, 7, 7, 5, 5, 7, 2, 6, 2, 4

Draw up a frequency table and hence draw a histogram of this data.

3)   The table below gives a grouped frequency distribution.

diameter (mm)	5.94–5.96	5.97–5.99	6.00–6.02	6.03–6.05	6.06–6.08
frequency	8	37	90	52	13

(a) Determine the class width.
(b) Draw a histogram for this distribution.

4)   For the grouped frequency distribution given below, draw a histogram and state the class width for each of the classes.

Resistance (ohms)	110–112	113–115	116–118	119–121	122–124
Frequency	2	8	15	9	3

ARITHMETIC MEAN

The arithmetic mean is found by adding up all the observations in a set and dividing the result by the number of observations. That is:

$$\text{Arithmetic mean} = \frac{\text{the sum of the observations}}{\text{the number of observations}}$$

EXAMPLE 1

Five turned bars are measured and their diameters were found to be: 15.03, 15.02, 15.02, 15.00 and 15.03 mm. What is their mean diameter?

$$\text{Mean diameter} = \frac{15.03 + 15.02 + 15.02 + 15.00 + 15.03}{5}$$

$$= \frac{75.10}{5} = 15.02 \text{ mm}$$

THE MEAN OF A FREQUENCY DISTRIBUTION

The mean of a frequency distribution must take into account the frequencies as well as the measured observations.

EXAMPLE 2

5 castings have a mass of 20.01 kg each, 3 have a mass of 19.98 kg each and 2 have a mass of 20.03 kg each. What is the mean mass of the 10 castings?

The total mass is $(5 \times 20.01) + (3 \times 19.98) + (2 \times 20.03) = 200.05$ kg.

$$\text{Mean mass} = \frac{\text{total mass of the castings}}{\text{number of castings}} = \frac{200.05}{10} = 20.005 \text{ kg}$$

If $x_1, x_2, x_3 \ldots x_n$ are measured observations which have frequencies $f_1, f_2, f_3 \ldots f_n$ then the mean of the distribution is:

$$\bar{x} = \frac{x_1 f_1 + x_2 f_2 + x_3 f_3 + \ldots + x_n f_n}{f_1 + f_2 + f_3 \ldots + f_n} = \frac{\Sigma xf}{\Sigma f}$$

The symbol Σ simply means the 'sum of'. Thus Σxf tells us to multiply together corresponding values of x and f and add the results together.

EXAMPLE 3

Find the mean of the frequency distribution shown below:

x	14.96	14.97	14.98	14.99	15.00	15.01	15.02	15.03	15.04
f	2	4	11	20	23	21	9	8	2

The best way of setting out the work is to make a table as shown over.

x	f	xf
14.96	2	29.92
14.97	4	59.88
14.98	11	164.78
14.99	20	299.80
15.00	23	345.00
15.01	21	315.21
15.02	9	135.18
15.03	8	120.24
15.04	2	30.08
	100	1500.09
	$= \Sigma f$	$= \Sigma xf$

$$\bar{x} = \frac{\Sigma xf}{\Sigma f} = \frac{1500.09}{100} = 15.000\,9$$

THE CODED METHOD FOR CALCULATING THE MEAN

The calculation of the mean may be speeded up considerably by using a unit method which is often referred to as using a coded method. The first step is to choose any value in the x column to use as a datum for determining the coded values. A column may then be drawn up containing the actual values of x in terms of units above or below the chosen value of x. The calculation for Example 3 would be as follows:

Chosen value of $x = 15.00$. Unit size $= 0.01$.

x	14.96	14.97	14.98	14.99	15.00	15.01	15.02	15.03	15.04
x_c	-4	-3	-2	-1	0	$+1$	$+2$	$+3$	$+4$

The coded value for $x = 14.96$ is -4 because this value of x is 4 units *less* than the chosen value of x. Similarly the coded value for $x = 15.02$ is $+2$ because 15.02 is 2 units *greater* than the chosen value of x. It is very important to assign to the coded value a plus or a minus sign depending on whether it is greater or less than the chosen value of x.

Although any value of x may be chosen as the datum, the arithmetic will be simpler if the middle value of x is chosen.

The mean may now be calculated from the coded values as follows:

x	x_c	f	$x_c f$
14.96	−4	2	−8
14.97	−3	4	−12
14.98	−2	11	−22
14.99	−1	20	−20
15.00	0	23	0
15.01	+1	21	21
15.02	+2	9	18
15.03	+3	8	24
15.04	+4	2	8
		100	+9
		$= \Sigma f$	$= \Sigma x_c f$

$$\bar{x}_c = \frac{\Sigma x_c f}{\Sigma f} = \frac{9}{100} = 0.09$$

Actual value of $\bar{x}$ = (chosen value of x)+($\bar{x}_c \times$ unit size)

$$= 15.00 + 0.09 \times 0.01 = 15.00 + 0.000\,9 = 15.000\,9$$

MEAN OF A GROUPED DISTRIBUTION

The mean of a grouped distribution is found by taking the value of x as the mid-points of the class intervals. Again, it is best to use the coded method as shown in Example 4.

EXAMPLE 4

Find the mean of the grouped distribution shown in the table below.

diameter (mm)	7.45–7.47	7.48–7.50	7.51–7.53	7.54–7.56	7.57–7.59
frequency	16	34	28	18	4

Diameter (mm)	x	x_c	f	$x_c f$
7.45–7.47	7.46	−6	16	−96
7.48–7.50	7.49	−3	34	−102
7.51–7.53	7.52	0	28	0
7.54–7.56	7.55	+3	18	+54
7.57–7.59	7.58	+6	4	+24
			100	−120

Chosen value of $x = 7.52$ mm

Unit size $= 0.01$ mm

$$x_c = \frac{-120}{100} = -1.2$$

Since $\bar{x}_c$ is negative it indicates that the mean is 1.2 units less than the chosen size. That is,

$$\bar{x} = 7.52 - (1.2 \times 0.01) = 7.52 - 0.012 = 7.508 \text{ mm}$$

Exercise 14

1)   The marks of a student in five examinations were as follows: 84, 90, 72, 60 and 74. Find his mean mark.

2)   The lengths of eight metal bars were measured with the following results: 109.23, 109.21, 108.98, 109.03, 108.98, 109.22, 109.20, 108.91 mm. What is the mean length of the bars?

3)   22 forgings have a mean mass of 12 kg and 18 similar forgings have a mean mass of 11.85 kg. What is the mean mass of the 40 forgings?

4)   A batch of 100 metal bars are made to a nominal diameter of 13.00 mm. On measuring the bars the following frequency distribution was obtained:

diameter (mm)	12.95	12.96	12.97	12.98	12.99	13.00	13.01
frequency	2	4	11	18	31	22	8

diameter (mm)	13.02	13.03	13.04
frequency	2	1	1

Calculate the mean length of the bars.

5)   A sample of 100 components was measured with the following results:

length (mm)	19.61	19.62	19.63	19.64	19.65	19.66	19.67	19.68	19.69
frequency	2	4	12	18	31	22	8	2	1

Calculate the mean lengths of the components.

6)   The table below shows the distribution of the maximum loads supported by certain cables.

max. load (kN)	19.2–19.5	19.6–19.9	20.0–20.3	20.4–20.7
frequency	4	12	18	3

Calculate the mean load which the cables will support.

7)   The table below shows a frequency distribution for the lifetime of cathode ray tubes.

lifetime (hours)	400–499	500–599	600–699	700–799	800–899
frequency	14	50	82	46	8

Calculate the mean lifetime of the cathode ray tubes.

8) The sales of steel ingots by the R. J. Steel Company Ltd. were as follows:

Mass of ingots (tonnes)	Number of ingots
1 and up to but less than 2	6
2 and up to but less than 3	17
3 and up to but less than 4	29
4 and up to but less than 5	32
5 and up to but less than 6	15
6 and up to but less than 7	1

Calculate the mean mass of the ingots sold.

THE MODE

The mode of a set of numbers is the number which occurs most frequently. Thus the mode of

$$2\ 3\ 3\ 4\ 4\ 4\ 5\ 5\ 6\ 6\ 7\ 8$$

is 4, since this number occurs three times which is more than any of the other numbers in the set.

For a set of numbers the mode may not exist. Thus the set of numbers

$$4\ 5\ 6\ 8\ 9\ 10\ 12$$

has no mode.

It is possible for there to be more than one mode. The set of numbers

$$2\ 3\ 3\ 5\ 5\ 5\ 6\ 6\ 7\ 8\ 8\ 8\ 9\ 10$$

has two modes, 5 and 8. The set of numbers is said to be *bimodal*. If there is only one mode, the set of numbers is said to be *unimodal*.

THE MODE OF A FREQUENCY DISTRIBUTION

The mode of a frequency distribution may be found by drawing a histogram as shown in Example 5.

EXAMPLE 5

The table below shows the distribution of maximum loads supported by certain cables produced by the Steel Wire Company. Draw a histogram of this information and hence find the mode of the distribution.

Maximum load (kN)	Number of cables
84–88	4
89–93	10
94–98	24
99–103	34
104–108	28
109–113	12
114–118	6
119–123	2

Assuming that the measurements of load were accurate only to the nearest kilonewton, the class boundaries are:

83.5–88.5, 88.5–93.5, 93.5–98.5, etc.

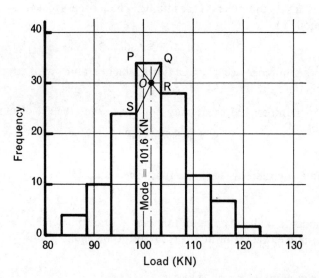

Fig. 3.3

The histogram is then as shown in Fig. 3.3. The mode is then found by drawing the diagonals PR and QS whose intersection is at O. The modal value is the value of the load corresponding to the point O. From the diagram this is found to be 101.6 kN.

Calculating the mode

The mode in Example 5 can be calculated by using the triangles OPS and OQR shown in Fig. 3.3. Fig. 3.4 shows the same triangles but to a larger scale.

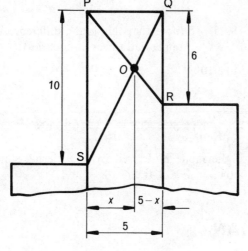

Fig. 3.4

Since triangles OPS and OQR are similar,

$$\frac{SP}{RQ} = \frac{x}{5-x}$$

$$\frac{10}{6} = \frac{x}{5-x}$$

$$10(5-x) = 6x$$

$$16x = 50$$

$$x = 3.1$$

The mode is found by adding 3.1 to the lower boundary for the class in which the mode occurs. Hence,

$$\text{mode} = 98.5 + 3.1 = 101.6 \text{ kN}$$

The mode can also be calculated by using the following formula:

$$\text{mode} = L + c\left(\frac{l}{l+u}\right)$$

where $L =$ lower boundary of the modal class,

$c =$ the class width,

$l =$ difference between the frequency of the modal class and the frequency of the next class below the modal class.

u = difference between the frequency of the modal class and the frequency of the next class above the modal class

Thus for the frequency distribution of Example 5.,

$L = 98.5$

$c = 98.5 - 93.5 = 5$ (note that this is the difference between the upper and lower boundaries)

$l = 34 - 24 = 10$

$u = 34 - 28 = 6$

$$\text{mode} = 98.5 + \left(\frac{10}{10+6}\right) \times 5 = 98.5 + \frac{50}{16} = 101.6 \text{ kN}$$

(This formula for the mode is derived by considering similar triangles similar to those shown in Fig. 3.4).

THE MEDIAN

If a set of numbers is arranged in ascending (or descending) order of size, the median is the value which lies half-way along the set. Thus for the set:

$$3, 4, 4, 5, 6, 7, 7, 9, 10$$

the median is 6.

If there is an even number of values the median is found by taking the average of the two middle values. Thus for the set:

$$3, \ 3, \ 5, \ 7, \ 9, \ 10, \ 13, \ 15$$

the median is $\frac{1}{2}(7+9) = 8$.

EXAMPLE 6

The hourly wages of five employees in an office are £1.52, £2.96, £2.28, £8.20 and £2.75.

Arranging the amounts in ascending order we have,

$$£1.52, \ £2.28, \ £2.75, \ £2.96, \ £8.20$$

The median is therefore £2.75,

Note that the median is not affected by the extreme amount £8.20. The mean, which is £3.542, is affected by it. In this case the median gives a better indication of the average hourly wage than does the mean.

DISCUSSION ON THE MEAN, MEDIAN AND MODE

The arithmetic mean is the most familiar kind of average and it is extensively used in statistical work. However, in some cases the mean is definitely misleading as shown in Example 6. Again, the mean size of screws used in a factory is not of much use to the purchasing officer. because it might be at some point between stock sizes. In such cases the mode is probably the best value to use. However, which average is used will depend upon the particular circumstances.

Exercise 15

1) Find the mode of the following set of numbers: 3, 5, 2, 7, 5, 8, 5, 2, 7, 6

2) Find the mode of: 38.7, 29.6, 32.1, 35.8, 43.2.

3) Find the modes of: 8, 4, 9, 3, 5, 3, 8, 5, 3, 8, 9, 5, 6, 7.

4) The data below relates to the resistance in ohms of an electrical part. Find the mode of this distribution, by drawing a histogram.

resistance (ohms)	119	120	121	122	123	124
frequency	5	9	19	25	18	4

5) Find the mode of the frequency distribution given in Question 7, Exercise 14.

6) The information below shows the distribution of the diameters of rivet heads for rivets manufactured by a certain company.

diameter (mm)	18.407–18.412	18.413–18.418	18.419–18.424
frequency	2	6	8

diameter (mm)	18.425–18.430	18.431–18.436	18.437–18.442
frequency	12	7	3

Find the mode of this distribution,

(a) by drawing a histogram
(b) by calculation.

7) Find the median of the following set of numbers: 9, 2, 7, 3, 8, 5, 4.

8) A student receives the following marks in an examination in five subjects: 84, 77, 95, 80 and 97. What is the median mark?

9) The following are the weekly wages earned by six people working in a small factory: £38, £71, £59, £63, £58 and £68. What is the median wage?

10)   Find the mean and the median for the following set of observations:
15.63, 14.95, 16.00, 12.04, 15.88 and 16.04 ohms. Which of the two, the
median or the mean, is, in your opinion, the better to use for these observations?

FREQUENCY CURVES

A frequency curve may be drawn by joining the mid-points of the top of
the rectangles in a histogram (Fig. 3.5). The frequency curve is a convenient
way of representing frequency distributions to make comparisons between
them easier (Fig. 3.6).

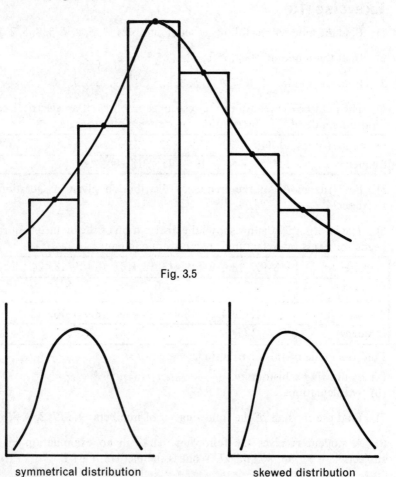

Fig. 3.5

symmetrical distribution   skewed distribution

Fig. 3.6

MEASURES OF DISPERSION

The central tendency of a distribution, as given by the mean, mode or median, gives some idea about the position of the distribution from the reference axis. For instance, Fig. 3.7 shows three similar distributions which have different means.

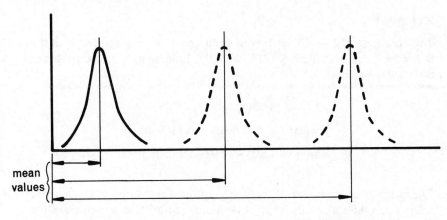

mean values

Fig. 3.7

However, Fig. 3.8 shows two different distributions which have the same mean but very different spread or dispersion. We need, therefore a measure which will define this spread or dispersion. The measures used are the range and the standard deviation.

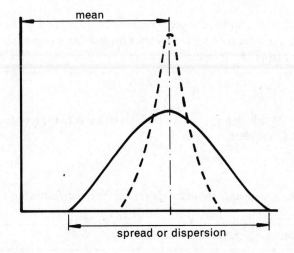

mean

spread or dispersion

Fig. 3.8

THE RANGE

The range is the difference between the largest observation in a set and the smallest observation in the set. That is,

$$\text{range} = \text{largest observation} - \text{smallest observation}$$

EXAMPLE 7

The diameters of five similar turned components were measured with the following results: 15.01, 15.03, 14.98, 14.99, 15.00 mm. Find the range for these measurements.

Smallest measurement $= 14.98$ mm

Largest measurement $= 15.03$ mm

Range $= 15.03 - 14.98 = 0.05$ mm

The range gives some idea of the spread of the distribution but it depends solely upon the end values. It gives no indication of the distribution of the data and hence it is never used as a measure of dispersion for a frequency distribution. However, if only a small number of observations are taken, as was the case in Example 7, the range is a very effective measure of dispersion.

THE STANDARD DEVIATION

The most valuable and widely used measure of dispersion is the standard deviation. It is always represented by the Greek letter σ (sigma) and it may be calculated from the formula,

$$\sigma = \sqrt{\frac{\Sigma x^2 f}{\Sigma f} - \bar{x}^2}$$

The best way of calculating the standard deviation is to use a coded method as shown in Example 8.

EXAMPLE 8

Calculate the mean and standard deviation for the following frequency distribution.

resistance (ohms)	5.37	5.38	5.39	5.40	5.41	5.42	5.43	5.44
frequency	4	10	14	24	34	18	10	6

Chosen value of $x = 5.40$ Unit size $= 0.01$ ohm

x	x_c	f	$x_c f$	$x_c^2 f$
5.37	-3	4	-12	36
5.38	-2	10	-20	40
5.39	-1	14	-14	14
5.40	0	24	0	0
5.41	1	34	34	34
5.42	2	18	36	72
5.43	3	10	30	90
5.44	4	6	24	96
		120	78	382

$$\bar{x}_c = \frac{\Sigma x_c f}{\Sigma f} = \frac{78}{120} = 0.65$$

$$\therefore \quad \bar{x} = 5.40 + 0.65 \times 0.01 = 5.406\ 5 \text{ ohm}$$

$$\sigma_c = \sqrt{\frac{\Sigma x_c^2 f}{\Sigma f} - (\bar{x}_c)^2} = \sqrt{\frac{382}{120} - (0.65)^2} = \sqrt{2.760\ 8} = 1.662$$

$$\therefore \quad \sigma = \sigma_c \times \text{unit size} = 1.662 \times 0.01 = 0.016\ 62 \text{ ohm.}$$

If the distribution is reasonably symmetrical, as is the one in Example 8, a rough check for the standard deviation may be obtained by finding the range of the data and dividing it by 6. Thus for Example 8,

$$\text{range} = 5.44 - 5.37 = 0.07$$

$$\sigma \text{ roughly} = \frac{0.07}{6} = 0.012 \text{ mm}$$

The calculated value of 0.016 57 is therefore of the right order (i.e. it is not wildly incorrect.)

The standard deviation for a grouped distribution may be calculated by taking x as the mid-points of the class widths as shown in Example 9.

EXAMPLE 9

Calculate the standard deviation for the distribution shown below.

length (mm)	15.23–15.25	15.26–15.28	15.29–15.31	15.32–15.34
frequency	2	5	7	12
length (mm)	15.35–15.37	15.38–15.40	15.41–15.43	15.44–15.46
frequency	17	9	5	3

Chosen value of $x = 15.36$ Unit size $= 0.03$ mm

Class	x	x_c	f	$x_c f$	$x_c{}^2 f$
15.23–15.25	15.24	-4	2	-8	32
15.26–15.28	15.27	-3	5	-15	45
15.29–15.31	15.30	-2	7	-14	28
15.32–15.34	15.33	-1	12	-12	12
15.35–15.37	15.36	0	17	0	0
15.38–15.40	15.39	1	9	9	9
15.41–15.43	15.42	2	5	10	20
15.44–15.46	15.45	3	3	9	27
			60	-21	173

A unit size of 0.03 mm has been chosen because each value of x differs from its preceding value by 0.03. Making the unit size as large as possible simplifies the calculation of the standard deviation.

$$\bar{x}_c = \frac{-21}{60} = -0.35$$

$$\bar{x} = 15.36 + 0.03 \times (-0.35) = 15.349\ 5 \text{ mm}$$

$$\sigma = \sqrt{\frac{173}{60} - (-0.35)^2} = \sqrt{2.760\ 8} = 1.662$$

$$= 1.662 \times 0.03 = 0.049\ 85 \text{ mm}$$

(Rough check for σ gives $\dfrac{15.46 - 15.23}{6} = 0.038$, which is of the same order as the value calculated above.)

Exercise 16

1) Calculate the mean and standard deviation for the following five observations: 16.01, 16.00, 15.98, 15.97 and 15.99 mm.

2) Calculate the mean and standard deviation for the following frequency distribution:

diameter (mm)	11.46	11.47	11.48	11.49	11.50	11.51	11.52	11.53
frquency	1	4	12	15	11	6	3	1

3) An article is being produced in quantity on a machine tool. To check the length of the article 100 of them are measured with the results given below:

length (mm)	22.36	22.37	22.38	22.39	22.40	22.41	22.42	22.43
frequency	1	4	9	24	30	26	5	1

Calculate the mean and standard deviation.

4) Find the mean and the standard deviation for the grouped frequency distribution given in Question 6, Exercise 15.

5) Calculate the standard deviation for the distribution of loads supported by cables, Question 6, Exercise 14.

SUMMARY

a) The frequency is the number of times an observation occurs.

b) A frequency distribution may be represented diagrammatically by a histogram, which consists of a set of rectangles. If the class widths are all the same then the heights of the rectangles may be taken to represent the frequencies.

c) The width of the rectangles of a histogram is found by subtracting the lower class boundary from the upper class boundary.

d) The arithmetic mean is found by using the formula:

$$\bar{x} = \frac{\text{the sum of the observations}}{\text{the number of observations}}$$

e) For a frequency distribution $\bar{x} = \dfrac{\Sigma xf}{\Sigma f}$ and this is best calculated by using a coded method.

f) The mean of a grouped distribution is found by taking x as being the mid-point of the class intervals. The formula given in **e)** can then be used.

g) The mode of a set of observations is the observation which occurs most frequently.

h) For a frequency distribution the mode can be found by drawing a histogram or by using the formula: mode $= L + c\left(\dfrac{1}{1+u}\right)$.

i) The median of a set of numbers is found by arranging the numbers in ascending or descending order. If there is an odd number of quantities the median is the middle value. If there is an even set of quantities then the median is found by taking the mean of the two middle values.

j) A frequency curve may be drawn by joining the mid-points of the tops of the rectangles in a histogram. Frequency curves are a useful way of comparing frequency distributions.

k) Two measures of dispersion are commonly used. They are the range and the standard deviation.

l) Range = largest observation − smallest observation.

m) The standard deviation is found by using the following formula:

$$\sigma_c = \sqrt{\frac{\Sigma x_c^2 f}{\Sigma f} - (\bar{x}_c)^2} \text{ where } x_c \text{ is the coded value of } x.$$

$$\sigma = \sigma_c \times \text{unit size.}$$

n) A rough check for the standard deviation may be found by dividing the range of the distribution by 6.

Self Test 3

1) A histogram is:

    a a measure of mass   **b** a type of horizontal bar chart

    c a history of weights and measures

    d a diagram representing a frequency distribution

2) The range of a distribution is:

    a the smallest observation   **b** the largest observation

    c the difference between the smallest observation and the largest observation

    d another name for the standard deviation

3) For the numbers 13, 18, 12, 11, 13, 19, 11, 16 and 11, the number 13 is:

    a the mean   **b** the median   **c** the mode   **d** the range

4) The mean of four numbers is 14. Three of the numbers are 4, 10 and 16. What is the fourth number?

    a 10   **b** 36   **c** 30   **d** 26

5) What is the median of the following numbers: 4, 12, 6, 6, 8, 14 and 20?

    a 6   **b** 8   **c** 10   **d** 70

6) Some steel bars are measured to the nearest millimetre. What is the upper limit of the class 130 mm but less than 150 mm?

    a 140 mm   **b** 149 mm   **c** 150.5 mm   **d** 150 mm

7) The diameters of some ball bearings are measured to the nearest 0.01 millimetre. The measurements are grouped in classes. What is the width of the class interval 18.02–18.04?

    a 0.02 mm   **b** 0.03 mm   **c** 0.04 mm   **d** 0.01 mm

8) The information below relates to the mass in grams of packets of chemical.

Mass	Frequency
20	1
21	2
22	4
23	5
24	4
25	1

What is the mean of this distribution?

    a 23   **b** 22.3   **c** 22.7   **d** 0.7

9) In Question 8, what is the mode?

    a 22   **b** 23   **c** 24   **d** $22\frac{2}{3}$

10) The standard deviation for the distribution given in Question 8 is:

    a 22.7   **b** 5   **c** 23   **d** 1.3

11) The weekly wage of 30 employees of a firm were analysed with the following results:

Wage £	Frequency
24–28	5
28–32	12
32–36	10
36–40	3

What is the mean weekly wage?

    a £29.48   **b** £1.48   **c** £31.48   **d** £30.37

12) In Question 11, what is the modal class?

    a 28–32   **b** 32–36   **c** 30   **d** 34

13) In Question 11, what is the standard deviation?

    a £0.87   **b** £3.49   **c** £16   **d** £17

14) The standard deviation for the numbers 2, 4, 6, 8, 10 is:

    a 0   **b** 2.4   **c** $\sqrt{8}$ **d** 8

15) Which of the following is a measure of location (i.e. central tendency)?

    a arnge   **b** standard deviation   **c** mode

    d frequency

16) Which of the following is a measure of dispersion?

    a mode   **b** range   **c** median   **d** frequency

GRAPHS

On reaching the end of this chapter you should be able to:-
1. *Choose suitable scales for drawing a graph.*
2. *Plot three points from coordinates determined from an equation of the type y = mx+c and draw a straight line through these points.*
3. *Plot cordinates from a set of experimental data which obeys a linear law.*
4. *Draw a straight line to fix the points in 3.*
5. *Calculate the gradient of a straight line using the tangent ratio.*
6. *Distinguish between lines having positive, negative and zero gradients.*
7. *Identify m as the gradient and c as the intercept on the y-axis.*
8. *Determine the law of a straight line.*
9. *Use the law to calculate other values.*
10. *Reduce non-linear physical laws of the form y = axⁿ (n = −1, ½ or 2) to a linear equation.*
11. *Plot the corresponding straight line graph to verify the law.*

AXES OF REFERENCE

To plot a graph we first draw two lines at right angles to each other (Fig. 4.1). These lines are called the axes of reference. The horizontal axis is often called the x-axis and the vertical axis the y-axis.

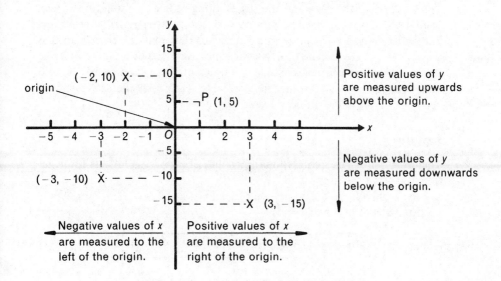

Fig. 4.1

CO-ORDINATES

Co-ordinates are used to mark the points on a graph. In Fig. 4.1 the point P has been plotted so that $x = 1$ and $y = 5$. The values 1 and 5 are said to be the rectangular co-ordinates of P. For brevity we say that P is the point $(1, 5)$.

In plotting graphs we may have to include co-ordinates which are positive and negative. To represent these on a graph we make use of the number scales used in directed numbers. As well as the point $(1, 5)$ the points $(3, -15)$, $(-2, 10)$ and $(-3, -10)$ are plotted in Fig. 4.1.

The Origin

If the zero of both axes occurs at the intersection of the axes as in Fig. 4.1, then this point $(0, 0)$ is called the *origin*.

AXES AND SCALES

The location of the axes and the scales along each axis should be chosen so that all the points may be plotted with the greatest possible accuracy. The scales should be as large as possible but they must be chosen so that they are easy to read. The most useful scales are 1, 2 and 5 units to 1 large square on the graph paper. Some multiples of these such as 10, 20, 100 units etc. per large square are also suitable. Note that the scales chosen need not be the same on both axes.

EXAMPLE 1

The table below gives corresponding values of x and y. Plot this information and from the graph find:
a) the value of y when $x = -3$
b) the value of x when $y = 2$

x		−4	−2	0	2	4	6
y		−2.0	−1.6	0	1.4	2.5	3.0

The graph is shown plotted in Fig. 4.2 and it is a smooth curve. This means that there is a definite law (or equation) connecting x and y. We can therefore use the graph to find corresponding values of x and y between those given in the original table of values. By using the constructions shown in Fig. 4.2:

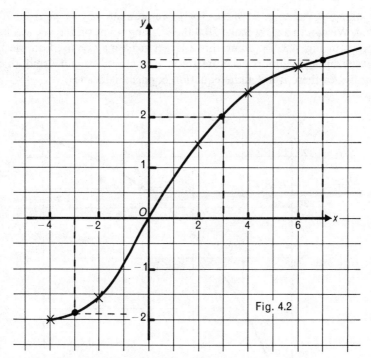

Fig. 4.2

a) the value of y is -1.9 when $x = -3$;
b) the value of x is 3 when $y = 2$.

Using a graph in this way to find values of x and y not given in the original table of values is called *interpolation*. If we extend the curve so that it follows the general trend we can estimate corresponding values of x and y which lie *just beyond* the range of the given values. Thus in Fig. 4.2 by extending the curve we can find the probable value of y when $x = 7$. This is found to be 3.2. Finding a probable value in this way is called *extrapolation*. An extrapolated value can usually be relied upon but in some cases it may contain a substantial amount of error. Extrapolated values must therefore be used with care.

It must be clearly understood that interpolation and extrapolation can only be used if the graph is a straight line or a smooth curve.

EXAMPLE 2

Corresponding values of x and y are shown in the table below

x	0	10	20	30	40	50
y	20.0	22.0	23.5	24.4	25.0	25.4

Illustrate this relationship on a graph.

Looking at the range of values for y, we see that they range from 20.0 to 25.4. We can therefore make 20.0 the starting point on the vertical axis as shown in Fig. 4.3. By doing this a larger scale may be used on the y-axis thus resulting in a more accurate graph. The graph is again a smooth curve and hence there is a definite equation connecting x and y.

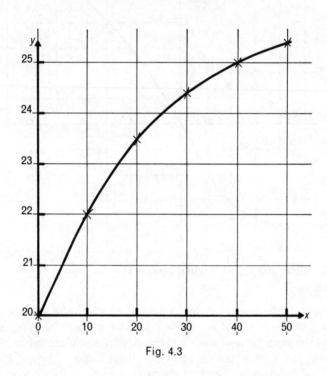

Fig. 4.3

GRAPHS OF SIMPLE EQUATIONS

Consider the equation:

$$y = 2x+5$$

We can give x any value we please and so calculate a corresponding value for y. Thus,

when $x = 0$ $y = 2 \times 0 + 5 = 5$
when $x = 1$ $y = 2 \times 1 + 5 = 7$
when $x = 2$ $y = 2 \times 2 + 5 = 9$ and so on

The value of y therefore depends on the value allocated to x. We therefore call y the *dependent variable*. Since we can give x any value we please, we call x the *independent variable*. It is usual to mark the values of the independent variable along the horizontal x-axis and the values of the dependent variable are then marked off along the vertical y-axis.

EXAMPLE 3

Draw the graph of $y = 2x - 5$ for values of x between -3 and 4.

Having decided on some values for x we calculate the corresponding values for y by substituting in the given equation. Thus,

$$\text{when } x = -3, y = 2 \times (-3) - 5 = -6 - 5 = -11$$

For convenience the calculations are tabulated as shown below.

x	-3	-2	-1	0	1	2	3	4
$2x$	-6	-4	-2	0	2	4	6	8
-5	-5	-5	-5	-5	-5	-5	-5	-5
$y = 2x - 5$	-11	-9	-7	-5	-3	-1	1	3

A graph may now be plotted using these values of x and y (Fig. 4.4). The graph is a straight line. Equations of the type $y = 2x - 5$, where the highest powers of the variables, x and y, is the first are called equations of the *first degree*. All equations of this type give graphs which are straight

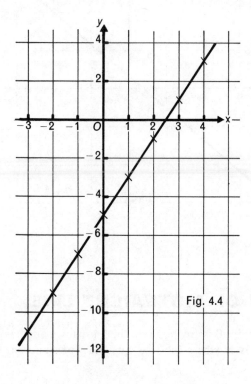

Fig. 4.4

lines and hence they are often called *linear equations*. In order to draw graphs of linear equations we need only take two points. It is safer, however, to take three points, the third point acting as a check on the other two.

EXAMPLE 4

By means of a graph show the relationship between x and y in the equation $y = 5x+3$. Plot the graph between $x = -3$ and $x = 3$.

Since this is a linear equation we need only take three points.

x	-3	0	$+3$
$y = 5x+3$	-12	3	$+18$

The graph is shown in Fig. 4.5.

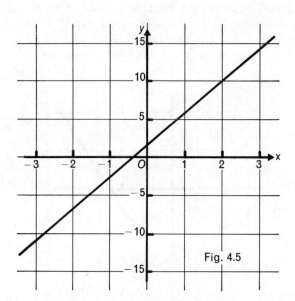

Fig. 4.5

THE LAW OF A STRAIGHT LINE

In Fig. 4.6, the point B is any point on the line shown and has co-ordinates x and y. Point A is where the line cuts the y-axis and has co-ordinates $x = 0$ and $y = $ c.

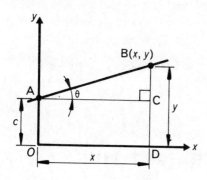

Fig. 4.6

In △ABC $$\frac{BC}{AC} = \tan\theta$$

∴ $$BC = (\tan\theta).AC$$

but also $$y = BC + CD$$

$$= (\tan\theta).AC + CD$$

∴ $$y = mx + c$$

where $$m = \tan\theta$$

and $$c = \text{distance } CD = \text{distance } OA$$

m is called the *gradient of the line*.

c is called the *intercept on the y-axis*. Care must be taken as this only applies if the origin (i.e. the point $(0, 0)$) is at the intersection of the axes.

In mathematics the gradient of a line is defined as the tangent of the angle that the line makes with the horizontal, and is denoted by the letter m.

Hence in Fig. 4.6 the gradient $= m = \tan\theta = \dfrac{BC}{AC}$

(Care should be taken not to confuse this with the gradient given on maps, railways, etc. which is the sine of the angle (not the tangent) — e.g. a railway slope of 1 in 100 is one unit vertically for every 100 units measured along the slope.)

Fig. 4.7 shows the difference between positive and negative gradients.

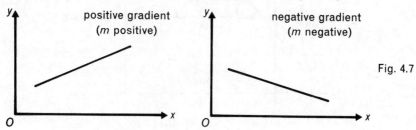

positive gradient
(*m* positive)

negative gradient
(*m* negative)

Fig. 4.7

Summarising:

The standard equation, or law, of a straight line is $y = mx + c$

where m is the gradient

and c is the intercept on the y-axis.

OBTAINING THE STRAIGHT LINE LAW OF A GRAPH

Two methods are used:

(i) Origin at the intersection of the axes

When it is convenient to arrange the origin, i.e. the point $(0, 0)$, at the intersection of the axes the values of gradient m and intercept c may be found directly from the graph as shown in Example 5.

EXAMPLE 5

Find the law of the straight line shown in Fig. 4.8.

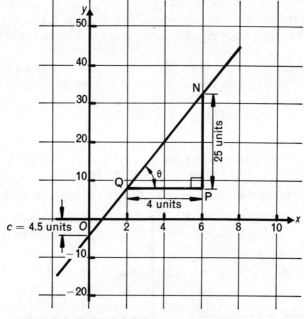

Fig. 4.8

To find gradient *m*. Take any two points Q and N on the line and construct the right angled triangle QPN. This triangle should be of reasonable size, since a small triangle will probably give an inaccurate result. Note that if we can measure to an accuracy of 1 mm using an ordinary rule, then this error in a length of 20 mm is much more significant than the same error in a length of 50 mm.

The lengths of NP and QP are then found using the scales of the *x* and *y* axes. Direct lengths of these lines as would be obtained using an ordinary rule, e.g. both in centimetres, must *not* be used — the scales of the axes must be taken into account.

$$\therefore \qquad \text{gradient } m = \tan \theta = \frac{NP}{QP} = \frac{25}{4} = 6.25$$

To find intercept *c*. This is measured again using the scale of the *y*-axis.

$$\therefore \text{ intercept} \qquad c = -4.5$$

The law of the straight line.

The standard equation is $\quad y = mx + c$

$\therefore$ the required equation is $\; y = 6.25x + (-4.5)$

i.e. $\qquad\qquad\qquad\qquad y = 6.25x - 4.5$

(ii) Origin not at the intersection of the axes

This method is applicable for all problems — it may be used, therefore, when the origin is at the intersection of the axes.

If a point lies on line then the co-ordinates of that point satisfy the equation of the line, e.g. the point (2, 7) lies on the line $y = 2x + 3$ because if $x = 2$ is substituted in the equation, $y = 2 \times 2 + 3 = 7$ which is the correct value of *y*. Two points, which lie on the given straight line, are chosen and their co-ordinates are substituted in the standard equation $y = mx + c$. The two equations which result are then solved simultaneously to find the values of *m* and *c*.

EXAMPLE 6

Determine the law of the straight line shown in Fig. 4.9.

Choose two convenient points P and Q and find their co-ordinates. Again these points should not be close together, but as far apart as conveniently possible. Their co-ordinates are as shown in Fig. 4.9.

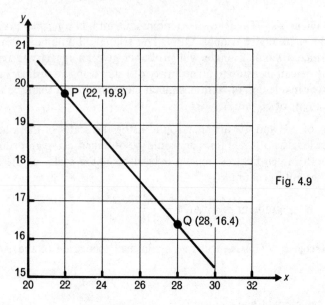

Fig. 4.9

Let the equation of the line be $y = mx+c$

Now P (22, 19.8) lies on the line $\therefore$ $19.8 = m(22)+c$

and Q (28, 16.4) lies on the line $\therefore$ $16.4 = m(28)+c$

To solve these two equations simultaneously we must first eliminate one of the unknowns. In this case c will disappear if the second equation is subtracted from the first, giving

$$19.8-16.4 = m(22-28)$$

i.e. $$3.4 = m(-6)$$

$\therefore$ $$m = \frac{3.4}{-6}$$

$\therefore$ $$m = -0.567$$

To find c the value of $m = -0.567$ may be substituted into either of the original equations. Choosing the first equation we get

$$19.8 = -0.567(22)+c$$

i.e. $$19.8 = -12.47+c$$

$\therefore$ $$c = 19.8+12.47$$

$\therefore$ $$c = 32.27$$

Hence the required law of the straight line is

$$y = -0.567x+32.27$$

GRAPHS OF EXPERIMENTAL DATA

Readings which are obtained as a result of an experiment will usually contain errors owing to inaccurate measurement and other experimental errors. If the points, when plotted, show a trend towards a straight line or a smooth curve this is usually accepted and the best straight line or curve drawn. In this case the line will not pass through some of the points and an attempt must be made to ensure an even spread of these points above and below the line or the curve.

One of the most important applications of the straight line law is the determination of a law connecting two quantities when values have been obtained from an experiment as Example 7 illustrates.

EXAMPLE 7

During a test to find how the power of a lathe varied with the depth of cut results were obtained as shown in the table. The speed and feed of the lathe were kept constant during the test.

depth of cut, d (mm)	0.51	1.02	1.52	2.03	2.54	3.0
power, P (W)	0.89	1.04	1.14	1.32	1.43	1.55

Show that the law connecting d and P is of the form $P = ad+b$ and find the law. Hence find the value of d when P is 1.2 watts.

The standard equation of a straight line is $y = mx+c$. It often happens that the variables are *not* x and y. In this example d is used instead of x and is plotted on the horizontal axis, and P is used instead of y and is plotted on the vertical axis.

Similarly the gradient $= a$ instead of m, and the intercept on the y-axis $= b$ instead of c.

On plotting the points (Fig. 4.10) it will be noticed that they deviate slightly from a straight line. Since the data are experimental we must expect errors in observation and measurement and hence a slight deviation from a straight line must be expected.

The points, therefore, approximately follow a straight line and we can say that the equation connecting P and d is of the form $P = ad+b$.

Because the origin is *not* at the intersection of the axes, to find the values of constants a and b we must choose two points *which lie on the line*. These two points must be as far apart as possible in order to obtain maximum accuracy.

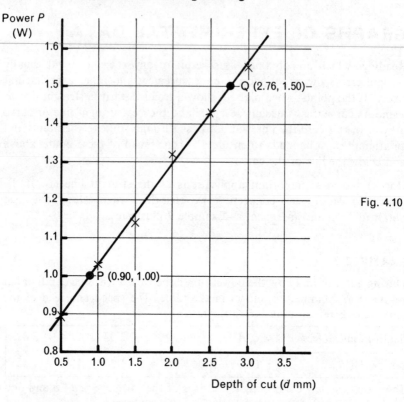

Fig. 4.10

In Fig. 4.10 the points P (0.90, 1.00) and Q (2.76, 1.50) have been chosen.

The point P(0.90, 1.00) lies on the line $\therefore$ $1.00 = a(0.90)+b$

and point Q(2.76, 1.50) lies on the line $\therefore$ $1.50 = a(2.76)+b$

Now subtracting the first equation from the second we get:

$$1.50-1.00 = a(2.76-0.90)$$

$\therefore$ $$0.50 = a(1.86)$$

$\therefore$ $$a = \frac{0.50}{1.86} = 0.27$$

Now substituting the value $a = 0.27$ into the first equation we get:

$$1.00 = 0.27(0.90)+b$$

$\therefore$ $$1.00 = 0.24+b$$

$\therefore$ $$b = 1.00-0.24$$

$\therefore$ $$b = 0.76$$

Hence the required law of the line is $P = 0.27d+0.76$

To find d when $P = 1.2$ W. The value $P = 1.2$ is substituted into the equation:

$$P = 0.27d + 0.76$$

giving

$$1.2 = 0.27d + 0.76$$

$$\therefore \qquad d = \frac{1.2 - 0.76}{0.27}$$

$$\therefore \qquad d = 1.63 \text{ mm} \quad \text{(since all values of } d \text{ are mm}$$
$$\text{when values of } P \text{ are watts.)}$$

This value of d may be verified by checking the corresponding value of d corresponding to $P = 1.2$ on the straight line in Fig. 4.10.

Any inaccuracies may be due to rounding off calculations to two significant figures, e.g. the value of m is $\dfrac{0.5}{1.86} = 0.269$ if three significant figures are considered. Bearing in mind, however, the experimental errors etc. the rounding off as shown seems reasonable. This question of accuracy is always open to debate, the most dangerous error being to give the calculated results to a far greater accuracy than the original given data.

Exercise 17

1) Draw the straight line which passes through the points $(4, 7)$ and $(-2, 1)$. Hence find the gradient of the line and its intercept on the y-axis.

2) The following equations represent straight lines. Sketch them and find in each case the gradient of the line and the intercept on the y-axis.

(a) $y = x + 3$ (b) $y = -3x + 4$ (c) $y = -3.1x - 1.7$
(d) $y = 4.3x - 2.5$

3) A straight line passes through the points $(-2, -3)$ and $(3, 7)$. *Without* drawing the line find the values of m and c in the equation $y = mx + c$.

4) The width of keyways for various shaft diameters are given in the table below.

dia. of shaft D (mm)	10	20	30	40	50	60	70	80
width of keyway W (mm)	3.75	6.25	8.75	11.25	13.75	16.25	18.75	21.25

Show that D and W are connected by a law of the type $W = aD + b$ and find the values of a and b.

5) During an experiment to find the coefficient of friction between two metallic surfaces the following results were obtained.

load W (N)	10	20	30	40	50	60	70
friction force F (N)	1.5	4.3	7.6	10.4	13.5	15.6	18.8

Show that F and W are connected by a law of the type $F = aW+b$ and find the values of a and b.

6)   In a test on a certain lifting machine it is found that an effort of 50 N will lift a load of 324 N and that an effort of 70 N will lift a load of 415 N. Assuming that the graph of effort plotted against load is a straight line find the probable load that will be lifted by an effort of 95 N.

7)   The following results were obtained from an experiment on a set of pulleys. W is the load raised and E is the effort applied. Plot these results and obtain the law connecting E and W.

W (N)	15	20	25	30	35	40	45
E (N)	2.3	2.7	3.2	3.8	4.3	4.7	5.3

8)   During a test with a thermo-couple pyrometer the e.m.f. (E millivolts) was measured against the temperature at the hot junction (t °C) and the following results obtained:

t	200	300	400	500	600	700	800	900	1 000
E	6	9.1	12.0	14.8	18.2	21.0	24.1	26.8	30.2

The law connecting t and E is supposed to be $E = at+b$. Test if this is so and find suitable values for a and b.

9)   The resistance (R ohms) of a field winding is measured at various temperatures (t °C) and the results recorded in the table below:

t(°C)	21	26	33	38	47	54	59	66	75
R (ohms)	109	111	114	116	120	123	125	128	132

If the law connecting R and t is of the form $R = a+bt$ find suitable values of a and b.

NON-LINEAR LAWS WHICH CAN BE REDUCED TO THE LINEAR FORM

Many non-linear equations can be reduced to the linear form by making a suitable substitution.

Common forms of non-linear equations are, a and b being constants:

$$y = \frac{a}{x} + b$$

$$y = \frac{a}{x^2} + b$$

$$y = ax^2 + b$$
$$y = a\sqrt{x} + b$$

Consider $y = \dfrac{a}{x} + b$

Let $z = \dfrac{1}{x}$ so that the equation becomes $y = az + b$. If we now plot values of y against the corresponding values of z we will get a straight line since $y = az + b$ is of the standard linear form. In effect y has been plotted against $\dfrac{1}{x}$.

The following example illustrates this method:

EXAMPLE 8

The voltage V across the arc of a carbon filament lamp for values of the current I in the arc were measured in an experiment and the results are shown in the following table:

I	1.0	1.5	2.0	2.5	3.0	3.5	4.0
V	82.0	68.7	62.0	58.0	55.3	53.4	52.0

The relation between V and I is thought to be of the form $V = \dfrac{a}{I} + b$. Check this and find the values of a and b.

By putting $z = \dfrac{1}{I}$ the equation will be reduced to $V = az + b$ which is a straight line equation. To try this we shall draw up a table of V and z, and plot the values obtained.

I	1.0	1.5	2.0	2.5	3.0	3.5	4.0
$z = \dfrac{1}{I}$	1.0	0.667	0.500	0.400	0.333	0.286	0.250
V	82.0	68.7	62.0	58.0	55.3	53.4	52.0

The graph obtained is shown in Fig. 4.11, and since it is a straight line it follows the law is of the form $V = az + b$, i.e. of the form $V = \dfrac{a}{I} + b$.

To find the values of a and b the two point method must be used since the *origin* is *not* at the intersection of the axes.

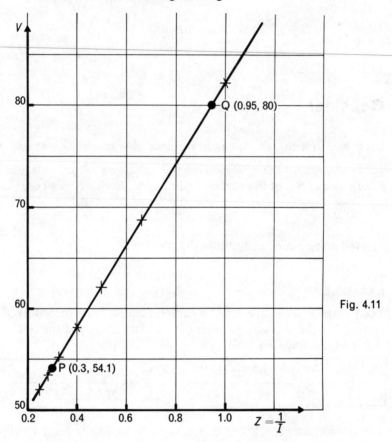

Fig. 4.11

The point (0.30, 54.1) lies on the line $\therefore\ 54.1 = a(0.30)+b$

and the point (0.95, 80.0) lies on the line $\therefore\ 80.0 = a(0.95)+b$

hence subtracting the first equation from the second we get,

$$80.0 - 54.1 = a(0.95 - 0.30)$$

i.e. $25.9 = a(0.65)$

$\therefore$ $a = \dfrac{25.9}{0.65} = 40$

Substituting this value in the first equation we obtain:

$$54.1 = 40(0.30)+b$$

from which $b = 54.1 - 12$

$\therefore$ $b = 42$

Hence the values of a and b are 40 and 42 respectively.

Consider $y = ax^2 + b$.

Let $z = x^2$ and as previously if we plot values of y against z (in effect x^2) we will get a straight line since $y = az + b$ is of the standard form. The following example illustrates this method:

EXAMPLE 9

The fusing current I amperes for wires of various diameters d mm is as shown below:

d (mm)	5	10	15	20	25
I (amperes)	6.25	10	16.25	25	36.25

It is suggested that the law $I = ad^2 + b$ is true for the range of values given, a and b being constants. By plotting a suitable graph show that this law holds and from the graph find the constants a and b. Using the values of these constants in the equation $I = ad^2 + b$ find the diameter of the wire required for a fusing current of 12 amperes.

By putting $z = d^2$ the equation $I = ad^2 + b$ becomes $I = az + b$ which is the standard form of a straight line. Hence by plotting I against d^2 we should get a straight line if the law is true. To try this we draw up a table showing corresponding values of I and d^2.

d	5	10	15	20	25
$z = d^2$	25	100	225	400	625
I	6.25	10	16.25	25	36.25

From the graph (Fig. 4.12) we see that the points do lie on a straight line and hence the values obey a law of the form $I = ad^2 + b$.

To find the values of a and b choose two points which lie on the line and find their co-ordinates.

The point (400, 25) lies on the line, $\therefore$ $25 = 400a + b$ (1)

The point (100, 10) lies on the line, $\therefore$ $10 = 100a + b$ (2)

Subtracting equation (2) from equation (1),

$$15 = 300a$$

$$a = 0.05$$

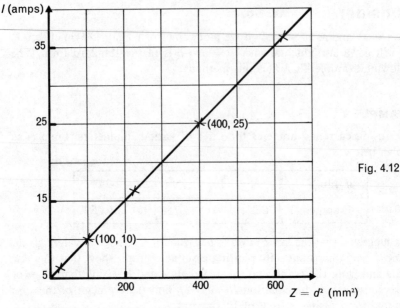

Fig. 4.12

Substituting $a = 0.05$ in equation (2),

$$10 = 100 \times 0.05 + b$$

$$b = 5$$

Therefore the law is:

$$I = 0.05d^2 + 5$$

When $I = 12$,

$$12 = 0.05d^2 + 5$$

$$0.05d^2 = 7$$

$$d^2 = \frac{7}{0.05} = 140$$

$$d = \sqrt{140} = 11.83 \text{ mm}$$

Consider $y = \dfrac{a}{x^2} + b.$

Let $z = \dfrac{1}{x^2}$ so that the equation becomes $y = az + b$. If we now plot values of y against corresponding values of z we will get a straight line since $y = az + b$ is of the standard linear form. In effect y has been plotted against $\dfrac{1}{x^2}$.

Consider $y = a\sqrt{x} + b.$

Let $z = \sqrt{x}$ and as previously if we plot values of y against z (in effect $\sqrt{x}$) we will get a straight line since $y = az + b$ is of the standard linear form.

SUMMARY

a) Co-ordinates are the values of x and y which fix a point on a graph with axes at right angles to each other — the point is denoted by (x, y).

b) The *origin* is the point $(0, 0)$.

c) When drawing a graph make full use of the space available on the sheet of graph paper.

d) Choose convenient scales, i.e. 1, 2, or 5 units (or these multiplied by powers of 10) per square of the graph paper.

e) A straight line graph may be called a linear graph. Two points only are necessary to fix a straight line, but a third point serves as a check.

f) The law of a straight line is $y = mx + c$ where m is the gradient and c is the intercept on the vertical axis (providing that the origin is at the intersection of the axes.)

g) The gradient m is the tangent of the angle the line makes with the horizontal.

positive gradient   negative gradient

h) If the origin is at the intersection of the axes then the gradient m may be found by drawing a suitable (i.e. reasonably large) triangle and dividing the vertical height by the horizontal length. The intercept c may be read directly from where the line cuts the vertical axis.

i) If the *origin* is *not* at the intersection of the axes then the two-point method must be used to determine m and c. (**N.B.** this method also holds if the origin is at the intersection of the axes).

j) The following equations represent graphs which may be reduced to a straight line form by making the corresponding substitutions:

$$y = \frac{a}{x} + b \qquad \text{substitute} \qquad z = \frac{1}{x}$$

$$y = \frac{a}{x^2} + b \qquad \text{substitute} \qquad z = \frac{1}{x^2}$$

$$y = ax^2 + b \qquad \text{substitute} \qquad z = x^2$$

$$y = a\sqrt{x} + b \qquad \text{substitute} \qquad z = \sqrt{x}$$

Exercise 18

1) Corresponding values obtained experimentally for two quantities are:

x	1	2	3	4	5
y	4.9	11.1	21.0	34.1	52.9

A law of the form $y = ax^2 + b$ is suspected. By plotting y vertically against x^2 horizontally show that the law is as stated. Hence find values for a and b.

2) The values tabulated below are thought to obey the law $y = m\sqrt{x} + c$. By plotting y (vertically) against $\sqrt{x}$ (horizontally) show that this is so and hence find values for m and c.

x	1	4	9	16	25
y	2	3.5	5	6.5	8

3) The values tabulated below follow the law $y = \frac{a}{x} + b$. By plotting y (vertically) against $\frac{1}{x}$ (horizontally) prove the law and find suitable values for a and b.

x	0.1	0.2	0.4	0.5	1.0
y	31	16	8.5	7	4

4) The resistance R N/tonne, to the motion of a train travelling at a speed of v km/h is shown in the table below. The law connecting R and v may be of the type $R = av^2 + b$. Test if this is so and hence find the values of a and b.

v	8	16	24	32	48	64
R	23.6	26.7	32.5	40.0	62.3	93.4

5) The table below shows how the coefficient of friction, μ, between a belt and a pulley varies with the speed, v m/s, of the belt. By plotting a graph show that $\mu = m\sqrt{v} + c$ and find the values of m and c.

μ	0.26	0.29	0.32	0.35	0.38
v	2.22	5.00	8.89	13.89	20.00

6) The following readings were taken during a test:

R (ohms)	85	73.3	64	58.8	55.8
I (amperes)	2	3	5	8	12

R and I are thought to be connected by an equation of the form $R = \dfrac{a}{I} + b$.

Verify that this is so by plotting R (y-axis) against $\dfrac{1}{I}$ (x-axis) and hence find values for a and b.

7) In an experiment, the resistance R of copper wire of various diameters d mm was measured and the following readings obtained.

d mm	0.1	0.2	0.3	0.4	0.5
R ohms	20	5	2.2	1.3	0.8

Show that $R = \dfrac{k}{d^2}$ and find a suitable value for k.

8) The fusing current for different diameters of a certain wire is as shown below.

diameter (x mm)	5	10	15	20	25
fusing current (I amperes)	6.25	10	16.25	25	36.25

It is thought that $I = ax^2 + b$. By plotting a suitable graph show that this is so and hence find suitable values for a and b.

Self-Test 4

State which answer or answers are correct:

1) The origin is the point:

    a (1, 1)   **b** (0, 1)   **c** (1, 0)   **d** (0,0)

2) The recommended number of units per square when choosing a scale for a graph are:

    a 3   **b** 2   **c** 40   **d** 10

3) A line with a negative gradient is:

    a   **b**   **c**   **d**

4) Which of the following points lie on the line $y = 2x + 3$?

    a (2, 1)   **b** (−2, 0)   **c** (2, 3)   **d** (1, 5)

5) Which of these equations represents a straight line?

    a $y = x^2 + 2$ **b** $y = 2x + 3$ **c** $y = 7 - x$ **d** $y^2 = x + 1$

6) The gradient of the line $y = 4x + 8$ is:

    a 8   **b** 1   **c** 2   **d** 4

7) The gradient of the line $y = 6 - 2x$ is:

    a 2   **b** 6   **c** -2 **d** -3

8) The values of gradient and intercept for the line $y = 3x$ are:

    a 1 and 3   **b** 3 and 1   **c** 1 and 0   **d** 3 and 0

9) The graph of $y = 2x + 3$ will look like one of the diagrams in Fig. 4.13.

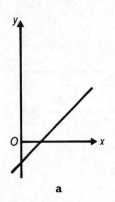 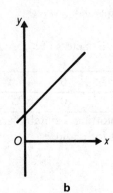 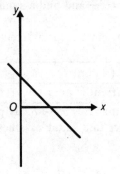 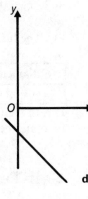

    a   **b**   **c**   **d**

Fig. 4.13

10) The graph of $y = 5 - 3x$ will look like one of the following diagrams (Fig. 4.14).

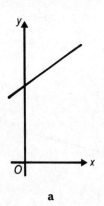 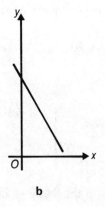 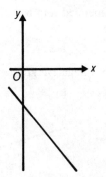 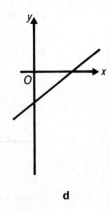

    a   **b**   **c**   **d**

Fig. 4.14

11) A straight line passes through the points (0, 1) and (2, 7). The law of the line is therefore:

a $y = 3x+1$ **b** $y = 3x-1$ **c** $y = \frac{3}{7}x+1$ **d** $y = \frac{3}{7}x-1$

12) The graph showing the relationship between two quantities S and T is a straight line. Values of S are indicated on the horizontal axis. The gradient of the graph is 5 and the intercept on the vertical axis is 3. Hence the law of the line is:

a $S = 5T+3$ **b** $T = 5S+3$ **c** $S = 3T+5$ **d** $T = 3S+5$

13) The law of the line shown in Fig. 4.15 is:

a $y = 2x+5$ **b** $y = 5-2x$ **c** $y = 2x-1$ **d** $y = \frac{1}{2}x+\frac{1}{2}$

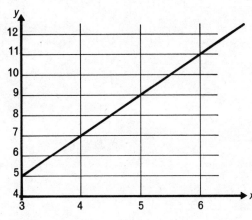

Fig. 4.15

14) The relation between two variables M and N is $M = \frac{a}{N}+b$. To obtain a straight line values of M are plotted against values of:

a N **b** $\frac{1}{N}$ **c** $\frac{a}{N}$ **d** MN

15) A straight line graph is obtained by plotting values of $\sqrt{P}$ horizontally and Q vertically. The relationship is given by:

a $\sqrt{P} = aQ+b$ **b** $P = a\sqrt{Q}+b$ **c** $Q = a+b\sqrt{P}$
d $Q = a\sqrt{P}+b$

ANSWERS

ANSWERS TO CHAPTER 1

Exercise 1

1) (a) 3.381   (b) 10.13   (c) 25.9 4
2) (a) 41° 49′   (b) 40° 47′   (c) 22° 23′
3) 21.6 cm   **4)** 7.47 m
5) 44° 44′, 44° 44′, 90° 32′
6) 10.3 cm
7) (a) 9.33   (b) 2.64   (c) 5.29
8) (a) 60° 42′   (b) 69° 20′   (c) 53° 19′
9) 66° 7′, 66° 7′, 47° 46′, 3.84 cm
10) 91° 56′, 8.75 cm
11) (a) 4.53 m   (b) 2.11 m
    (c) 2.40 m   (d) 5.65 m
12) (a) 4.35   (b) 9.28   (c) 4.43
13) (a) 59° 2′   (b) 15° 57′   (c) 22° 20′
14) 7.70 cm
15) 2.86 m   **16)** 2.09 cm
17) (a) 12.86 m   (b) 15.32 m
18) 11.3 km, 45° W of S   **19)** 9.65 m
20) 137.9 m
21) 21° 48′, 16° 52′, 10.77 m

Exercise 2

1) (a) 1.573   (b) 1.083   (c) 1.019
    (d) 1.538   (e) 1.288   (f) 0.283 3
    (g) 0.023 9   (h) 0.848 6   (i) 0.008 7
    (j) 0.058 4   (k) 0.074 3   (l) 0.475 9
2) (a) 47° 40′   (b) 57° 19′   (c) 63° 49′
3) (a) 8.007   (b) 6.145   (c) 9.396
    (d) 13.65   (e) 7.437   (f) 18.43
4) (a) 30° 55′   (b) 23° 10′   (c) 63° 42′
5) 76.03 mm   **6)** 42° 54′   **7)** 6.833 cm
9) 3.631 cm   **10)** 108.2 mm

Exercise 4

1) (a) 0.857 2, −0.515 0, −1.664 3
    (b) 0.028 2, −0.999 6, −0.028 2
    (c) 0.976 4, −0.216 2, −4.516 9
    (d) −0.515 0, −0.857 2, 0.600 9
    (e) −0.859 7, −0.510 8, 1.683 1
    (f) −0.979 8, −0.200 0, 4.900 6
    (g) −0.663 3, 0.748 4, −0.886 3
    (h) −0.888 7, 0.458 4, −1.938 9

2) (a) 1.166 6, −1.941 6, −0.600 9
    (b) 35.445 4, −1.000 4, −35.431 3
    (c) 1.024 2, −4.626 3, −0.221 4
    (d) −1.941 6, −1.166 6, 1.664 3
    (e) −1.163 2, −1.957 7, 0.594 2
    (f) −1.020 6, −5.001 6, 0.204 1
    (g) −1.507 7, 1.336 2, −1.128 3
    (h) −1.125 2, 2.181 5, −0.515 8
3) 3.024 8   **4)** −0.28   **5)** $-\frac{4}{5}, -\frac{3}{4}$
6) 8° 14′, 171° 46′   **7)** 153° 13′, 206° 47′
8) (a) 45° 32′, 134° 28′ (b) 118° 46′
    (c) 43° 28′   (d) 119° 33′
9) (a) 33° 25′, 146° 35′ (b) 122° 6′
    (c) 118° 31′
10) 14° 34′; 165° 26′
11) 105° 42′; 254° 18′

Exercise 5

1) (a) 0.615 7, 0.781 3
    (b) 0.955 1, 0.309 0
    (c) 0.615 7, −0.788 0
    (d) 0.955 1, −0.309 0
    (e) −0.342 0, −0.939 7
    (f) −0.939 7, −0.342 0
    (g) −0.819 2, 0.573 6
    (h) −0.529 9, 0.848 0
2) 30°, 150°; −2.82   **3)** 78°, 282°
4) 23°, 157°

Exercise 6

1) (a) ∠ C = 71°, b = 5.906 cm
    c = 9.986 cm
    (b) ∠ A = 48° a = 71.52 mm
    c = 84.16 mm
    (c) ∠ B = 56° a = 3.741 m
    b = 9.528 m
    (d) ∠ B = 46° b = 13.60 cm
    c = 5.84 cm
    (e) ∠ C = 67° a = 1.508 m
    c = 2.361 m
    (f) ∠ C = 63° 32′ a = 9.486 mm
    b = 11.56 mm
    (g) ∠ B = 135° 38′ a = 9.393 cm
    c = 14.44 cm

(h) $\angle B = 81°\,54'$ $b = 9.947$ m
   $c = 3.609$ m

(i) $\angle A = 53°\,39'$ $a = 2124$ mm
   $b = 2\,390$ mm

(j) $\angle A = 45°\,30'$ or $134°\,30'$,
   $\angle B = 95°\,30'$ or $6°\,30'$
   $c = 23.72$ or 2.699 cm

(k) $\angle A = 13°\,51'$ $\angle B = 144°\,2'$
   $b = 17.16$ m

(l) $\angle A = 86°\,1'$ or $15°\,43'$
   $\angle B = 54°\,51'$ or $125°\,9'$
   $a = 112.2$ or 30.48 mm

(m) $\angle A = 44°\,46'$ $\angle C = 49°\,57'$
   $a = 10.69$ cm

(n) $\angle B = 93°\,49'$ $\angle C = 36°\,52'$
   $b = 30.26$ cm

(o) $\angle B = 48°\,31'$ $\angle C = 26°\,25'$
   $c = 4.247$ cm

2) (a) $c = 10.15$ cm $\angle A = 50°\,11'$
   $\angle B = 69°\,49'$

(b) $a = 11.81$ cm $\angle B = 44°\,42'$
   $c = 79°\,18'$

(c) $b = 4.989$ m $\angle A = 82°\,24'$
   $\angle C = 60°\,18'$

(d) $\angle A = 38°\,12'$ $\angle B = 81°\,38'$
   $\angle C = 60°\,10'$

(e) $\angle A = 24°\,42'$ $\angle B = 44°\,54'$
   $\angle C = 110°\,24'$

(f) $\angle A = 34°\,42'$ $\angle B = 18°\,6'$
   $\angle C = 127°\,12'$

3) 64.00 mm   4) $37°\,35'$

5) 14.2 cm, $142°\,39'$ 6) 6.02, 3.29 cm

7) $41°\,27'$ 8) $102°\,40'$

9) $104°\,28'$ 10) 14.5 cm

Exercise 7

1) 22.1 cm^2 2) 31.7 cm^2

3) 2765 mm each side   4) 540 cm^2

5) 738 mm^2

6) (a) 7.55 cm^2 (b) 8.06 m^2

7) 13.4 cm^2 8) 9.62 cm^2

9) (a) 143 cm^2 (b) 53.7 m^2 (c) 43.6 cm^2

10) 11.9 cm^2 11) 28 cm^2

12) 89 cm^2 13) 2.12 m

14) 3060 mm^2

15) (a) 13.8 cm^2 (b) 65 cm^2

16) (a) 3.31 mm^2 (b) 19.3 mm^2

17) 15.7 cm

Self Test 1

1) c   2) b   3) c   4) a

5) c, d   6) b   7) c, d   8) b

9) d   10) a, d   11) c   12) b

13) c   14) c   15) c   16) a, c

17) b, c   18) b, d   19) a,c,e   20) b, d

21) b   22) d   23) c, d

ANSWERS TO CHAPTER 2

Exercise 8

1) 20 mm   2) 3 m^2

3) 1400 cm^2 4) 630

Exercise 9

1) 25 mm   2) 333.333 m

3) 2.37 g/cm^3 4) 400 kg

5) 20 cm^3 6) 1500 ℓ

7) 1200 kg   8) 500 s

9) (a) 0.06 m^3/s   (b) 3600 ℓ/min

10) 1 m^2 11) 2 cm/min

Exercise 10

1) 8.8 mm   2) $0.012\,8$ m^2

3) (a) 1200 mm^2 (b) 276 mm^2

   (c) 261 mm^2 (d) 774 mm^2

   (e) 1050 mm^2 (f) 1094 mm^2

4) 28 cm^2 5) 89 cm^2

6) 2.12 m   7) 3060 mm^2

8) (a) 13.8 cm^2 (b) 65 cm^2

9) (a) 3.31 mm^2 (b) 19.3 mm^2

10) 15.7 cm   11) 2.99 cm

12) (a) $11\,200$ mm^2 (b) 3.02 cm^2

13) 6166 mm^2

14) (a) 22.0 mm   (b) 86.8 m   (c) 26.4 cm

15) (a) 10.94 mm   (b) 5.900 cm

   (c) 62.1 m

16) 6.2 cm^2 17) 3.41 cm

18) 2592.3 mm^2 19) 909

20) (a) 1.045 cm   (b) 2.29 cm

21) (a) $119.6°$ (b) $10.16°$

22) 8.92 cm

23) (a) 4.71 m^2 (b) 5.08 cm^2

   (c) 76.2 cm^2

24) 866 mm^2 25) 2.93 cm; 8.07 cm

26) 1239 mm^2 27) 1.85 cm^2

28) 163 mm^2

Exercise 11

1) 76.19 m   2) 10.61 m

3) $0.008\,75$ m^3 4) 36.7 cm

5) 2475 m   6) $128\,300$ mm^3

7) 6.542 cm   8) 140.3 mm

9) 75.4 mm   10) 1.77×10^6

11) 4157 cm³   **12)** 925 g
13) (a) 0.259 8 m³   (b) 0.002 16 m³
14) (a) 47.1 cm³, 83.3 cm²
   (b) 8.19 cm³, 19.6 cm²
15) 925 g
16) $V = t(\pi/4\ D^2 - l^2)$; 2.46 cm³
17) 50 265 ℓ **18)** 5.35 mm
19) 5.33 cm³, 20.5 cm²
20) 116 000 mm³, 14 900 mm²
21) 77.4 cm²
22) 30.9 cm, 32 720 cm², 4391 cm²
23) 19.9 ℓ, 3480 cm²
24) 20 cm   **25)** 15 250 cm³, 6218 cm²

Exercise 12

1) 752   **2)** 172   **3)** 0.8   **4)** 99
5) 25.5   **6)** 1095 m²   **7)** 17.2 m³
8) 1057 m², 9.46 m/s
9) approximately 3 000 000 m³

Self Test 2

1) d	**2)** b	**3)** d	**4)** d
5) b	**6)** a, c	**7)** b, d	**8)** b
9) c	**10)** b, d	**11)** d	**12)** a
13) a	**14)** b	**15)** d	**16)** b
17) a, d	**18)** a	**19)** d	**20)** c, d
21) a	**22)** c	**23)** b	**24)** a, c
25) b, d	**26)** a	**27)** d	**28)** a, c
29) d	**30)** a	**31)** b	**32)** c
33) c			

ANSWERS TO CHAPTER 3

Exercise 13

1)
diameter	24.95	24.96	24.97	24.98
frequency	1	1	4	4
diameter	24.99	25.00	25.01	25.02
frequency	5	5	7	6
diameter	25.03	25.04	25.05	
frequency	4	2	1	

2)
Length	0	1	2	3	4	5
Frequency	1	4	12	7	9	14
Length	6	7	8	9		
Frequency	13	7	9	4		

3) (a) 0.03 mm   **4)** 3 ohm

Exercise 14

1) 76   **2)** 109.095 mm
3) 11.932 5 kg   **4)** 12.989 3 mm

5) 19.648 6 mm   **6)** 19.966 kN
7) 641.5 hours   **8)** 3.86 tonnes

Exercise 15

1) 5   **2)** no mode
3) 3, 5 and 8   **4)** 121.96 ohms
5) 646.6 hours   **6)** 18.427 2 mm
7) 5   **8)** 84   **9)** £61
10) mean = 15.09 ohms,
    median = 15.755 ohms

Exercise 16

1) $\bar{x} = 15.99$, $\sigma = 0.0141\,4$
2) $\bar{x} = 11.492\,5$, $\sigma = 0.014\,52$
3) $\bar{x} = 22.398\,1$, $\sigma = 0.0127\,8$
4) $\bar{x} = 18.425\,4$, $\sigma = 0.007\,82$
5) $\bar{x} = 19.966\,2$, $\sigma = 0.316\,7$

Self Test 3

1) d	**2)** c	**3)** b	**4)** d
5) b	**6)** d	**7)** b	**8)** c
9) b	**10)** d	**11)** c	**12)** a
13) b	**14)** c	**15)** c	**16)** b

ANSWERS TO CHAPTER 4

Exercise 17

1) $m = 1, c = 3$
2) (a) $m = 1, c = 3$ (b) $m = -3, c = 4$
   (c) $m = -3.1, c = -1.7$
   (d) $m = 4.3, c = -2.5$
3) $m = 2, c = 1$
4) $a = 0.25, b = 1.25$
5) $a = 0.29, b = -1.0$
6) 529 N
7) $E = 0.098\,4W + 0.72$
8) $a = 0.03, b = 0$
9) $a = 100, b = 0.43$

Exercise 18

1) $a = 2, b = 3$ **2)** $m = 1.5, c = 0.5$
3) $a = 3, b = 1$ **4)** $a = 0.017\,4, b = 22$
5) $m = 0.040\,4, c = 0.20$
6) $a = 70, b = 50$ **7)** $k = 0.2$
8) $a = 0.05, b = 5$

Self Test 4

1) d	**2)** b, d	**3)** b	**4)** d
5) b, c	**6)** d	**7)** c	**8)** d
9) b	**10)** b	**11)** a	**12)** b
13) c	**14)** b	**15)** c, d	

INDEX